U.S. Department
of Transportation

**National Highway
Traffic Safety
Administration**

I0488963

DOT HS 810 685

NHTSA Technical Report

August 2006

An Evaluation of the 1998-1999 Redesign of Frontal Air Bags

1. Report No. DOT HS 810 685	2. Government Accession No.	3. Recipient's Catalog No.
4. Title and Subtitle AN EVALUATION OF THE 1998-1999 REDESIGN OF FRONTAL AIR BAGS		5. Report Date August 2006
		6. Performing Organization Code
7. Author(s) Charles J. Kahane, Ph.D.		8. Performing Organization Report No.
9. Performing Organization Name and Address Evaluation Division; National Center for Statistics and Analysis National Highway Traffic Safety Administration Washington, DC 20590		10. Work Unit No. (TRAIS)
		11. Contract or Grant No.
12. Sponsoring Agency Name and Address Department of Transportation National Highway Traffic Safety Administration Washington, DC 20590		13. Type of Report and Period Covered NHTSA Technical Report
		14. Sponsoring Agency Code

15. Supplementary Notes

16. Abstract

The first generation of frontal air bags saved the lives of thousands of drivers and adult or teenage right-front passengers. But they harmed occupants positioned close to the air bag at the time of deployment, especially infants and children. In 1998-1999, air bags were redesigned by depowering – by removing some of the gas-generating propellant or stored gas from their inflators – and/or by reducing the volume or rearward extent of air bags, positioning them further from occupants, tethering and hybrid inflators. NHTSA facilitated the redesign by permitting a sled test in lieu of a barrier impact to certify that air bags would protect an unrestrained occupant ("sled certification"). Statistical analyses of crash data through 2004 from NHTSA's Fatality Analysis Reporting System (FARS) and the Special Crash Investigations (SCI) compare fatality risk with sled-certified and first-generation air bags.

- The overall fatality risk in frontal crashes of 0-12 year-old child passengers in the front seat is a statistically significant 45 percent lower with sled-certified air bags than with first-generation air bags; fatalities caused by air bags in low-speed crashes were reduced by 83 percent.

- The overall fatality risk of drivers and of right-front passengers age 13 and older in frontal crashes is not significantly different with sled-certified air bags than with first-generation air bags; sled-certified air bags preserved the life-saving benefits of first-generation air bags.

17. Key Words NHTSA; FARS; Special Crash Investigations; FMVSS; statistical analysis; evaluation; benefits; effectiveness; fatality reduction; crashworthiness; depowering; sled certification	18. Distribution Statement Document is available to the public at the Docket Management System of the U.S. Department of Transportation, http://dms.dot.gov , Docket Number 26486.	
19. Security Classif. (Of this report) Unclassified	20. Security Classif. (Of this page) Unclassified	21. No. of Pages 97	22. Price

Form DOT F 1700.7 (8-72)   Reproduction of completed page authorized

TABLE OF CONTENTS

LIST OF ABBREVIATIONS

AAMA	American Automobile Manufacturers Association
AIS	Abbreviated Injury Scale
CATMOD	Categorical models procedure in SAS
CDS	Crashworthiness Data System (a sample of crashes in the United States since 1979)
CY	Calendar Year
DC	District of Columbia, United States
DOT	United States Department of Transportation
ESC	Electronic Stability Control
FARS	Fatality Analysis Reporting System (a census of fatal crashes in the United States since 1975)
FMVSS	Federal Motor Vehicle Safety Standard
HIC	Head Injury Criterion
IARV	Injury Assessment Reference Value
IR	Information Request (from NHTSA)
kPa	kilopascal
LTV	Light trucks and vans (includes pickup trucks, SUVs, minivans and full-sized vans)
MAIS	Maximum value, on the Abbreviated Injury Scale, of the injuries sustained by a person
mph	miles per hour
msec	millisecond
MY	Model year
NHTSA	National Highway Traffic Safety Administration
NPRM	Notice of Proposed Rulemaking

NVPP	R.L. Polk's National Vehicle Population Profile
RF	Right-Front
SAE	Society of Automotive Engineers
SAS	Statistical analysis software produced by SAS Institute, Inc.
SCI	Special Crash Investigations by NHTSA's National Center for Statistics and Analysis
SRP	Seating Reference Point
SUV	Sport utility vehicle
VIN	Vehicle Identification Number

EXECUTIVE SUMMARY

Frontal air bags saved the lives of an estimated 16,905 drivers and adult or teenage right-front passengers from 1987 through the end of 2004. But they can harm occupants positioned close to the air bag at the time of deployment. The National Highway Traffic Safety Administration (NHTSA) had long warned consumers not to place an infant's rear-facing safety seat in the front seat of a vehicle equipped with passenger air bags, because the infant would be close to the air bag at all times.

By late 1995, it was evident that not only infants, but also children and even some adults were injured by air bags; worse, statistical analyses showed a significant increase in overall fatality risk with air bags for children age 0-12 years (but a significant net benefit for adults and teenagers). Children slid forward toward the air bag during pre-crash braking, especially if they were unrestrained or on the lap of another passenger. Short drivers often sat close to the air bag located in the steering wheel. Any occupant could approach an air bag by leaning forward, for instance, to adjust the radio or air-conditioning. By late 1995, NHTSA had identified 30 fatalities due to contact with air bags in otherwise survivable crashes (that number had grown to 264 by the end of 2005).

Therefore, beginning in October 1995, NHTSA initiated immediate, interim and long-term actions to reduce and eventually eliminate the adverse effect of air bags for infants, children and other high-risk occupants while retaining, to the largest extent possible, the great life-saving benefits of air bags for most people:

- An immediate campaign to inform motorists of dangers from air bags, urging that children travel in the back seat. On-off switches for passenger air bags were factory-installed in pickup trucks without back seats. Subsequently, NHTSA also advised drivers to sit at least 10 inches away from the air bag. Drivers at high risk or people who must transport high-risk individuals in the front passenger seat could obtain aftermarket on-off switches at their own expense.

- An interim effort to furnish, as soon as possible, air bags that deploy less forcefully. On March 19, 1997, NHTSA modified the frontal impact test for the unrestrained dummy in Federal Motor Vehicle Safety Standard (FMVSS) No. 208 – "Occupant Crash Protection." That facilitated an industry-wide changeover, using already available technologies, to "redesigned" air bags in the very next model years (1998-1999).

- A long-term effort to develop "advanced" air bags that do not deploy at all for children ("suppression"), deploy only at a low level of force ("low-risk deployment"), or track an occupant's motion and suppress the air bag if they are too close ("dynamic automatic suppression"). Advanced air bags were phased into the fleet starting September 1, 2003 and were required in all new vehicles by September 1, 2006.

The interim effort – the redesign of frontal air bags in 1998-1999 – is the focus of this report. In the 1980s, the vast majority of fatalities were unrestrained, and the first generation of air bags was designed to protect unrestrained occupants as well as restrained occupants in frontal crashes. Air bags had to pass two 30 mph crash tests, where the vehicle itself impacts a rigid barrier, with unrestrained dummies in the driver's and right-front passenger's seats on one of the tests and restrained dummies on the other. Air bags had to fill up quickly with enough gas to cushion a large, unrestrained occupant, resulting in some designs having high initial pressure and velocity – and high injury risk for any person who might have strayed into the first two or three inches of the deployment path. By 1995-1996, serious injuries to children needed immediate remedy, whereas the great increase in belt use, with buckle-up laws in almost every State, had shifted priority away from the unrestrained occupant.

An imminently available technology to reduce the initial pressure and velocity of deployments was to "**depower**" air bags by removing some of the gas-generating propellant or stored gas from their inflators. Other innovations already in progress included reducing the volume or rearward extent of air bags, positioning them further from occupants, revised folding techniques, tethering and shifting from pyrotechnic inflators to hybrids including stored gas. The **"redesign"** of air bags in 1998-1999 consisted of depowering and/or some of the other innovations.

The manufacturers, however, doubted that depowered air bags would readily meet the existing FMVSS No. 208 crash test with unbelted dummies at 30 mph into a rigid barrier. The American Automobile Manufacturers Association proposed that the 30 mph unbelted crash test be optionally replaced with a sled test, also at 30 mph, with a standardized crash pulse lasting 125 milliseconds, a substantially more gradual deceleration than the barrier test on a typical vehicle. NHTSA and the safety community agreed that the proposal would facilitate the immediate redesign of air bags. NHTSA began to allow the sled test as an option on March 19, 1997. Make-models became "**sled-certified**" when the manufacturer chose the sled test option (whether or not the air bag was actually redesigned). Make-models comprising approximately 70 percent of sales were sled-certified for the entire 1998 model year. By the beginning of the 1999 model year, about 99 percent were sled-certified.

In other words, not every sled-certified air bag is necessarily depowered or redesigned, and vice-versa. But the overlap is great. For the vehicles where NHTSA has full information about air bag performance, 84 percent of the driver air bags were depowered upon sled certification and 2 percent, although not depowered, had some other feature substantially redesigned. On the passenger side, 70 percent were depowered upon sled certification and 11 percent were substantially redesigned in some other way, most often by shifting from pyrotechnic to hybrid inflators.

Because NHTSA knows when almost every make-model was sled-certified, but has only partial information on depowering and other redesign, the analyses in this report primarily compare the fatality risk in **sled-certified** vehicles to the risk in the model years just before sled certification.

Before it allowed sled certification, NHTSA tested the performance of depowered air bags and analyzed data to quantify the likely impact on safety:

- *Out-of-position occupants*: Based on deployment tests with dummies located close to the air bag, NHTSA was confident that depowering would prevent a lot of the fatalities with first-generation air bags. The agency anticipated a 44 percent reduction of fatalities caused by air bags to children age 1-12 years, and a large but unquantifiable reduction for out-of-position drivers and adult passengers. The agency did not estimate any specific benefit from depowering for infants in rear-facing seats.

- *Correctly positioned drivers and adult passengers*:
    - o *Unrestrained*: Substantial gas pressure is needed to cushion an unbelted occupant's thorax. Depowering, intuitively, could allow some unrestrained adults to "punch through" the air bag and be more severely injured by the structures beneath it, such as the steering assembly. Tests, however, did not always show increased risk. The agency anticipated a fatality increase ranging from near zero to possibly as high as 16 percent.
    - o *Belted*: "Punch-through" is less likely because safety belts absorb much of the occupant's kinetic energy, and a less aggressive air bag might help in certain crashes. Tests did not show a clear directional effect. The agency anticipated little net change in fatality risk, ranging from a 2 percent reduction to a 3 percent increase.

This report analyzes crash data up to calendar year 2004 to find what actually happened to fatality risk after air bags were sled-certified. NHTSA's Special Crash Investigations (SCI) identify the fatalities to infants, children and out-of-position adults due to contact with air bags in otherwise survivable, low-to-moderate speed crashes (defined as less than 25 mph Delta V): "SCI fatalities." The Fatality Analysis Reporting System (FARS), a census of fatal crashes, provided data to compare overall fatality risk in frontal crashes before and after sled certification.

The results are quite positive. Sled-certified air bags greatly reduced SCI fatalities to infants, children and adults, even beyond NHTSA's expectations. They did not completely eliminate that risk. Children up to and including age 12 years should continue riding in the back seat to avoid harm from air bags. In pickup trucks without a back seat but with an on-off switch available for the passenger air bag, that switch should be turned off when a child passenger up to and including age 12 years is riding in the front seat.

Sled-certified air bags entirely preserved the great life-saving benefits of first-generation air bags for belted drivers and for passengers age 13 years and up. Even for unbelted drivers and passengers, overall fatality risk did not change significantly relative to first-generation air bags.

In short, the combination of redesigned air bags and campaigns urging that children travel in the back seat reduced the SCI fatality rate for infants and children by 93 percent without any significant overall side effect for adults. It has been a highly successful interim measure until advanced air bags could be developed and placed in service.

The findings of the statistical analyses are the following:

THE SAFETY PROBLEM – IN PERSPECTIVE

- Before sled certification, a large percentage of the passenger fatalities in frontal crashes among children up to age 10 years were caused by air bags in low-to-moderate speed crashes and would have been survivable without the air bag ("SCI fatalities"). Here is a table of the occupant fatalities at seat positions equipped with first-generation air bags (not sled-certified) in calendar years 1990-2003:

	SCI Fatals	All Frontal Fatals	% SCI
Child passengers (age 0-12)	**146**	**351**	**41.60**
Less than 1 year	27	42	64.29
1 – 5 years	80	158	50.63
6 – 10 years	37	117	31.62
11 – 12 years	2	34	5.88
Adult passengers (age 13+)	**9**	**3,899**	**.23**
Drivers	**78**	**29,361**	**.27**
Unbelted females ≤ 5'3", age 70+	7	180	3.88

- Less than 1 percent of fatalities to drivers and to passengers age 13 years and up were SCI fatalities.

- Even for the most vulnerable group of drivers – unrestrained females up to 5'3" tall and at least 70 years old – only 4 percent of frontal fatals were SCI cases, about the same as for pre-teen passengers age 11-12 years.

MANUFACTURER RESPONSE TO SLED CERTIFICATION

- The extent of depowering may be described by the percent reduction in two parameters:
  - The peak pressure of the deploying air bag
  - The average rise rate of the pressure as it approaches its peak.

- The peak pressure of air bags decreased by an average of 13 percent on the driver's side upon sled certification, and likewise by 13 percent on the passenger side.

- The rise rate decreased by an average of 24 percent on the driver's side upon sled certification, and by 18 percent on the passenger side.

EFFECT OF SLED-CERTIFICATION ON SCI FATALITIES OF CHILDREN

- Here is a table of child passenger fatalities age 0-12 years caused by air bags in low-to-moderate speed crashes that would have been survivable without the air bag ("SCI fatalities") – per billion vehicle years, during calendar years 1998-2003. The year-to-year changes in children's front-seat occupancy are not a factor because the table compares SCI fatality rates during the same calendar years:

Calendar Years	Vehicles	SCI Fatalities	Vehicle Years	SCI Fatalities per Billion Years
1998-2003	Not sled-certified	75	213,481,330	353.0
1998-2003	Sled-certified	15	243,523,549	61.6

- For all children age 0-12 years in the front seat, the fatality rate in sled-certified vehicles was 83 percent lower than in the vehicles with first-generation air bags, during the same time period (calendar years 1998-2003).

- During 1998-2003 there were 11 SCI fatalities to infants in rear-facing safety seats in vehicles with first-generation air bags, **none** with sled-certified air bags.

- For 0-5 year-old children not in rear-facing seats, the SCI fatality rate in calendar years 1998-2003 was 78 percent lower with sled-certified air bags than with first-generation air bags; for 6-10 year-old children, 82 percent lower.

- Actual SCI fatality reductions for children exceed the 44 percent anticipated by NHTSA.

- During 1998-2003 there were 11 SCI fatalities to children age 0-12 years in vans with first-generation air bags, **none** with sled-certified air bags.

- For 0-12 year-old children, the SCI fatality rate in calendar years 1998-2003 was reduced with sled-certified air bags by 78 percent in passenger cars and by 93 percent in SUVs.

EFFECT OF SLED-CERTIFICATION ON SCI FATALITIES OF DRIVERS

- Drivers' SCI fatality rate in calendar years 1998-2003 was 70 percent lower with sled-certified air bags than with first-generation air bags.

EFFECT OF SLED-CERTIFICATION ON SCI FATALITIES OF ADULT AND TEENAGE PASSENGERS

- The SCI fatality rate of right-front passengers age 13 years and up in calendar years 1998-2003 was 42 percent lower with sled-certified air bags than with first-generation air bags.

COMBINED EFFECT OF SLED-CERTIFICATION AND MOVING TO THE BACK SEAT ON CHILDREN'S SCI FATALITIES

- For all children age 0-12 years, the fatality rate in sled-certified vehicles during calendar years 1998-2003 was 93 percent lower than it had been in vehicles with first-generation air bags, during calendar years 1990-1997.

Calendar Years	Vehicles	SCI Fatalities	Vehicle Years	SCI Fatalities per Billion Years
1990-1997	Not sled-certified	65	73,305,874	886.7
1998-2003	Sled-certified	15	243,523,549	61.6

- That reduction is the combined effect of safer air bags and behavioral changes over time: moving children to the back seat, and if children stayed in the front seat, at least moving the seat all the way back and increasing use of restraints.

EFFECT OF SLED-CERTIFICATION ON CHILDREN'S OVERALL FATALITY RISK IN FRONTAL IMPACTS

- The fatality risk in frontal crashes of 0-12 year-old child passengers in the front seat is a statistically significant 45 percent lower with sled-certified air bags than with first-generation air bags (90 percent confidence bounds: 30 to 56 percent).

- Fatality reductions were substantial for every type of vehicle:

Vehicle Type	Frontal Fatality Reduction (%) with Sled-Certified Air Bags
Passenger car	30
Pickup truck	55
SUV	48
Van	89

- Fatality reductions were larger for the younger children:

Age of the Child Passenger	Frontal Fatality Reduction (%) with Sled-Certified Air Bags
Less than 1 year	59
1 – 5 years	58
6 – 10 years	31
11 – 12 years	8

- The results are consistent with the great reduction of SCI fatalities with sled-certified air bags. For children up to age 10 years, approximately half the fatalities in frontal crashes with original air bags were SCI fatalities.

EFFECT OF SLED-CERTIFICATION ON DRIVERS' OVERALL FATALITY RISK IN FRONTAL IMPACTS

- The overall fatality risk of drivers in frontal crashes is the same with sled-certified air bags as with first-generation air bags (95 percent confidence bounds range from a 5 percent reduction to a 4 percent increase in fatality risk with sled-certified air bags).

- The fatality risk of **belted** drivers in frontal crashes is 5 percent lower with sled-certified air bags than with first-generation air bags. The reduction is not statistically significant.

- The fatality risk of **unrestrained** drivers in frontal crashes is 5 percent higher with sled-certified air bags than with first-generation air bags. The increase is not statistically significant.

- Sled-certified air bags preserved the life-saving benefits of first-generation air bags for drivers.

EFFECT OF SLED-CERTIFICATION ON ADULT AND TEENAGE PASSENGERS' OVERALL FATALITY RISK IN FRONTAL IMPACTS

- The overall fatality risk of right-front passengers age 13 years and older in frontal crashes is 5 percent lower with sled-certified air bags than with first-generation air bags. The reduction is not statistically significant (95 percent confidence bounds range from a 13 percent reduction to a 5 percent increase in fatality risk with sled-certified air bags).

- The fatality risk of **belted** right-front passengers in frontal crashes is 5 percent lower with sled-certified air bags than with first-generation air bags. The reduction is not statistically significant.

- The fatality risk of **unrestrained** right-front passengers in frontal crashes is 7 percent lower with sled-certified air bags than with first-generation air bags. The reduction is not statistically significant.

- Sled-certified air bags preserved the life-saving benefits of first-generation air bags for front-seat passengers age 13 years and up.

CHAPTER 1

THE REDESIGN OF FRONTAL AIR BAGS IN 1998-1999

1.1   The need to redesign frontal air bags

Frontal air bags for drivers and right-front passengers have great net benefits. The National Highway Traffic Safety Administration (NHTSA) estimates they saved 16,905 lives from 1987 through the end of 2004.[1] Nevertheless, air bags, especially up to the mid-1990s presented risks to occupants positioned close to the air bag at the time of deployment. The potential for air bags to harm at least some very nearby occupants is intuitively evident and had been corroborated by laboratory simulations during the 1970s.[2] Analyses in the 1970s and 1980s attempted to compare the risk to expected benefits, given the estimates of air bag effectiveness in those days. The analyses encouraged optimism that air bags would result in far greater benefits than harm to almost every subgroup of occupants – except infants.[3]

An infant in a rear-facing safety seat, placed in the front seat of a vehicle, is close to the passenger air bag at all times. In 1991, when passenger air bags began to appear in substantial numbers, NHTSA warned consumers that rear-facing safety seats should never be used in the front seat of a vehicle equipped with air bags.[4] On September 2, 1993, the agency amended Federal Motor Vehicle Safety Standard (FMVSS) No. 208 to require vehicles with passenger air bags manufactured after August 31, 1994 to have warning labels that rear-facing safety seats should never be placed in the front seat (and that other occupants should not sit or lean close to the air bag). The labels were placed on sun visors and in vehicle owners' manuals. Similar labels were required on child safety seats.[5] On May 23, 1995, NHTSA amended FMVSS No. 208 to permit on-off switches for the passenger air bag in pickup trucks without back seats or

---

[1] *Traffic Safety Facts – 2004 Data – Occupant Protection*, NHTSA Report No. DOT HS 809 909, Washington, 2005.

[2] Patrick, L.M. and Nyquist, G.W., *Airbag Effects on the Out-of-Position Child*, Paper No. 720442, Society of Automotive Engineers, Warrendale, PA, 1972; Aldman B, Anderson A, and Saxmark O., "Possible Effects of Air Bag Inflation on a Standing Child," *Proceedings of the 18th Annual Conference of the American Association for Automotive Medicine*, Morton Grove, IL, 1974.

[3] For example, *Automobile Occupant Crash Protection Progress Report No. 3*, NHTSA Report No. DOT HS 805 474, Washington, 1980, pp. 71-75 analyzes potential risk and benefits and concludes that "On balance, air bags will provide substantial crash protection to small children in crashes." See Kahane, C.J., *Lives Saved by the Federal Motor Vehicle Safety Standards and Other Vehicle Safety Technologies, 1960-2002*, NHTSA Technical Report No. DOT HS 809 833, Washington, 2004, p. 108 for a summary of early predictions of air bag effectiveness.

[4] *NHTSA Warns Parents about Child Safety Seat Use in Cars with Air Bags*, Press Release No. NHTSA 60-91, U. S. Department of Transportation, Office of the Assistant Secretary for Public Affairs, Washington, 1991.

[5] *Federal Register* 58 (September 2, 1993): 46551.

other vehicles that cannot accommodate child safety seats in the back seat. The amendment went into effect on June 22, 1995.[6]

By late 1995, field experience with air bags had demonstrated that not only infants in rear-facing seats, but also children and even some adults were injured by air bags. Children could slide forward toward the air bag during pre-crash braking, especially if they were unrestrained or on the lap of another passenger. Short drivers often sat close to the air bag located in the steering wheel. Any occupant could approach an air bag by leaning forward, for instance, to adjust the radio or air-conditioning. By the end of 1995, NHTSA's Special Crash Investigation (SCI) program had identified 30 people with fatal injury from contact with air bags in otherwise survivable crashes (change in vehicle velocity, Delta V < 25 mph), including 3 infants in rear-facing seats; 10 children, not in rear-facing seats, ranging from 4 to 9 years old; and 17 drivers, 10 of them 5'2" or shorter.[7] At the same time, statistical analyses of the Fatality Analysis Reporting System (FARS) did not show early air bags having life-saving benefits for children age 0-12 years greater than their harm; on the contrary, children's overall fatality risk significantly increased with air bags up to approximately age 10 years.[8]

Therefore, beginning in October 1995 NHTSA initiated a series of immediate, mid-term and long-term actions to reduce and eventually eliminate the adverse effect of air bags for infants, children and other high-risk occupants while retaining, to the largest extent possible, the great life-saving benefits of air bags for most people:

- On October 27, 1995 the agency launched a campaign to inform motorists of the dangers of air bags to children as well as infants, urging that children travel in the back seat whenever possible. On May 21, 1996 the campaign expanded to a government/industry coalition for air bag safety. The percentage of children riding in the front seat has greatly decreased over the years (see Section 1.5).[9]

- On November 9, 1995, NHTSA issued a request for comments and announced a public meeting to discuss technological changes to reduce adverse effects of air bags. The notice appraised several technologies. On August 6, 1996, the agency followed up with a Notice of Proposed Rulemaking (NPRM) to reduce adverse effects of air bags. Potential areas for regulation included labeling requirements, test procedures to facilitate redesign of air bags, and new, advanced air bag technologies.[10]

---

[6] *Federal Register* 60 (May 23, 1995): 27233; *Code of Federal Regulations*, Title 49, Government Printing Office, Washington, 2005, Part 571.208 S4.5.4. See also: *NHTSA Permits Air Bag Switch to Prevent Injury to Infants in Rear-Facing Safety Seats*, Press Release No. NHTSA 30-95, U. S. Department of Transportation, Office of the Assistant Secretary for Public Affairs, Washington, 1995; the press release, dated May 18, 1995, still says "air bags offer excellent supplemental protection to … most children" but "pose a unique safety risk to infants in rear-facing child seats."

[7] See also: Healy, J.R., and O'Donnell, J., "Deadly Air Bags," *USA Today*, July 8, 1996.

[8] Kahane, C.J., *Fatality Reduction by Air Bags*, NHTSA Technical Report No. DOT HS 808 470, Washington, 1996, pp. 44-49.

[9] *Safety Agency Issues Warning on Air Bag Danger to Children*, Press Release No. NHTSA 72-95, U. S. Department of Transportation, Office of the Assistant Secretary for Public Affairs, Washington, 1995; *Secretary Peña Announces Government/Industry Coalition for Air Bag Safety*, Press Release No. NHTSA 24-96, U. S. Department of Transportation, Office of the Assistant Secretary for Public Affairs, Washington, 1996.

[10] *Federal Register* 60 (November 9, 1995): 56554, 61 (August 6, 1996): 40784.

- On November 27, 1996, the agency fortified the warning labels, effective February 25, 1997, to explicitly warn of air bags' risk to children up to age 12 years and to specify that the back seat is the safest place for children.[11]

- On March 19, 1997, NHTSA amended FMVSS No. 208, effective immediately, relaxing some aspects of the frontal impact test for the unrestrained dummy in order to facilitate the introduction of "redesigned" air bags that deploy less forcefully. Redesigned air bags were introduced in the 1998 models, or not long thereafter.[12]

- On November 21, 1997, NHTSA enabled people who must transport high-risk individuals in the front seats of any vehicle to obtain aftermarket on-off switches at their own expense, starting on January 19, 1998. The agency also advised the public that drivers should sit at least 10 inches away from the air bag.[13]

- On May 12, 2000, the agency amended FMVSS No. 208 to phase in "advanced" air bags from September 1, 2003 to September 1, 2006. Advanced air bags do not deploy at all for children ("suppression"), deploy only at a low level of force ("low-risk deployment"), or track an occupant's motion and suppress the air bag if they are too close ("dynamic automatic suppression").[14]

The fourth action, the "redesign" of frontal air bags in 1998-1999 is the focus of this report. We will show that redesigned air bags greatly reduced the risk of a fatal injury, due to contact with the air bag, for a child passenger or a driver in a low-speed crash. They did not completely eliminate that risk. They did not significantly change (for better or worse) the effectiveness of air bags for belted drivers and adult passengers in higher severity crashes.

Despite these actions, fatal injuries continued to occur, primarily in the many pre-1998 vehicles that were still on the road. By the end of 2005, NHTSA had records of 264 fatalities due to contact with air bags in otherwise survivable crashes, including 24 infants in rear-facing safety seats; 142 children, not in rear-facing seats, ranging from 4 months to 11 years old; 85 drivers; and 13 adult passengers.

1.2   How frontal air bags work – and why they can pose a risk

A brief review of the purpose and technology of air bags illustrates why they pose a risk to some occupants. A frontal air bag system includes sensors at various locations in the vehicle that send an electrical signal if they experience a substantial deceleration and velocity change from a frontal direction. A control module commands the air bags to deploy if these signals imply that the vehicle has been in a relatively severe frontal impact. Traditionally, air bag assemblies include a charge of Sodium Azide propellant that will generate Nitrogen gas upon firing and/or a cylinder of compressed Argon gas and an inflatable bag made of fabric. Many air bags today use proprietary chemical compounds as propellants. All bags have vents or porous fabric to release the gas gradually after a deployment. Many have tethers (internal straps) that limit how far the

---

[11] *Federal Register* 61 (November 27, 1996): 60206.

[12] *Federal Register* 62 (March 19, 1997): 12960; *Code of Federal Regulations*, Title 49, Part 571.208 S13.

[13] *Federal Register* 62 (November 21, 1997): 62406; *Code of Federal Regulations*, Title 49, Part 595; *Air Bags & On-Off Switches*, NHTSA Publication No. DOT HS 808 629, Washington, 1997.

[14] *Federal Register* 65 (May 12, 2000): 30679; *Code of Federal Regulations*, Title 49, Part 571.208 S14.

bag can deploy toward the occupant, and/or internal baffles to make it spread outward, over a larger area. These assemblies are located in the steering wheel hub for the driver, and somewhere in the instrument panel for the passenger.

In a severe frontal crash such as a 30 mph barrier impact, even while the front bumper comes to an immediate stop against the barrier and the sheet metal deforms, the occupants remain in their seats for about the first 50 milliseconds or more, and the compartment interior continues forward at close to 30 mph as if nothing had happened yet. In the next 50-75 milliseconds, the compartment is slowed to a stop while unrestrained occupants continue to move forward at close to 30 mph, fly out of their seats, and strike the steering assembly, instrument panel and other structures at a high relative speed.

The technological marvel of the air bag is that it fully deploys in less than 50 milliseconds, before a correctly positioned occupant has even begun to move out of the seat. In that time, the sensors detect the crash, the control module sends signals to fire the propellant or open the cylinders, and gas is generated or released at high enough pressure for the bags to burst through "doors" in the steering wheel hub and instrument panel, and to fill up entirely. As the occupant begins to move forward, he or she will almost immediately contact an energy-absorbing surface that is ideally "tuned": soft enough to cushion the head and neck without serious injury, yet rigid enough to absorb much of the torso's kinetic energy. This energy is absorbed as the occupant compresses the bag and pushes the gas out through the vents.

The air bag is one key component in a chain of deformable devices that allow the occupant to "ride down" from 30 mph to 0 mph as gradually as possible, remaining in an upright position. The other components are: the continuing crumple of the sheet metal in the front of the vehicle even after the occupant moves out of the seat; the compression of the energy-absorbing steering assembly (FMVSS Nos. 203 and 204), facilitated by transmission of force through the air bag; the deformation of the instrument panel by the knees (FMVSS No. 201); and the "give" in the safety belt system (especially with load limiters) and the seat structure.

The safety belt is indeed vital for absorbing part of the occupant's kinetic energy and retarding forward motion. Take the safety belt away by failing to buckle up, and the air bag has to work substantially harder and perhaps sooner.

Under any circumstances, an air bag must deploy rapidly and with great force if it is to get its job done in less than 50 milliseconds. For adequate gas pressure (energy-absorbing capability) in the fully deployed bag, there must be much higher gas pressure and temperature as the bag begins to deploy – especially if the bag is designed to cushion an unbelted occupant. In the early 1980s, when belt use was less than 15 percent, protecting the unrestrained occupant was a priority.[15] Whereas air bags are of great value to the correctly positioned occupant, they can increase risk for an out-of-position occupant who is touching a deployment "door," or is within about 2 or 3 inches of it, at the moment of deployment.[16]

How do people get close to the air bag? As discussed above, infants in rear-facing child safety seats are close to the deployment door at all times, and must **never** be placed in the front seat of

---

[15] Kahane (2004), pp. 88-92.
[16] *Air Bags & On-Off Switches.*

a vehicle with a working passenger air bag. But risk is increased if the vehicle is designed for the air bag to emerge from the part of the instrument panel closest to the safety seat.

The majority of crashes that might require an air bag to deploy involve pre-impact braking.[17] During braking, a vehicle decelerates at rates up to about 1 g. Thus, an occupant is pulled forward relative to the vehicle interior with a force that can be equal to their own weight. In addition, prior to the major impact, minor impact can occur, moving the occupant towards the front of the vehicle. People may also be jostled forward from their seats by bumps on or off the road. Unrestrained children 1-12 years old are especially vulnerable, the younger the worse. They lack the inertial weight that keeps adults in their seats during the bumps and bounces and most of them are unable to support or steady themselves by having their feet on the floor or their hands on the instrument panel.

A safety belt or forward-facing safety seat can reduce the tendency to slide forwards but they are not panaceas. Especially if incorrectly used (for example, by routing the shoulder belt behind the back), they may allow considerable freedom of motion during normal driving and forward excursion upon deceleration. Drivers of short stature (usually less than 5 feet tall) sometimes sit within 10 inches of the steering wheel in order for their feet to reach the pedals. That makes them more prone than others to be thrown within 2 or 3 inches of the deployment door during the pre-crash sequence. Temporary proximity to the air bag can result from leaning forward to adjust the radio or other controls. In all cases, risk can increase if the air bag itself is stored in a part of the instrument panel close to the occupant or protrudes past the rim of the steering wheel close to the driver.

1.3   Quantitative analysis of risk from pre-1998 air bags

NHTSA's Special Crash Investigation (SCI) program identifies cases of fatalities due to occupant contact with air bags, as evidenced by detailed analysis of medical records and the vehicle interior, in otherwise survivable crashes, as evidenced by a Delta V estimated to be less than 25 mph. The Fatality Analysis Reporting System (FARS), on the other hand, is a census of all fatalities, including every occupant killed in a frontal crash who had been sitting at a position equipped with an air bag.[18] "Frontal" crashes are those with initial or principal impact at the 11, 12 or 1:00 location; the VIN identifies which vehicles are equipped with air bags. In the vast majority of the FARS cases, the air bag did not cause a fatality in an otherwise survivable crash. The crash was too severe for the air bag to prevent a fatality, taking into account the occupant's age and physical condition and/or the fatal injury resulted from sources or mechanisms that the air bag did not mitigate.

The ratio of SCI cases to frontal FARS cases, in vehicles before the redesign of 1998-1999 indicates the contribution of fatalities caused by air bags to the overall fatality picture – i.e., what percentage of fatalities was essentially caused by the air bag and probably would not have happened without the air bag. In turn, it will help us foresee whether a successful remedy to

---

[17] Montalvo, F., Bryant, R., and Mertz, H., *Possible Positions and Postures of Unrestrained Front-Seat Children at the Instant of Collision*, Paper No. 826045, Society of Automotive Engineers, Warrendale, PA, 1982.

[18] All front-seat passengers are included in vehicles with dual air bags – namely, right-front, center-front, sitting on somebody else's lap, other/unknown front-seat position.

reduce fatalities caused by air bags will substantially impact the overall fatality rate – or just be a drop in the bucket relative to overall rates. (Strictly speaking, not every SCI case is also a FARS frontal. Some incidents took place on private property such as parking lots and are not reportable to FARS; a few involved air bags deployments in primarily non-frontal impacts.)

Table 1-1 presents the ratios of SCI to FARS frontal fatalities for child passengers (age 0-12 years), adult passengers (age 13+ years) and drivers, for occupants at seat positions equipped with frontal air bags in model year 1989-1997 vehicles – i.e., before FMVSS No. 208 was modified to facilitate the redesign of air bags – in crashes that occurred during calendar years 1990-2003, reported to SCI, at least partially investigated, and included in the public SCI files by the end of 2005. (SCI notification and investigation ought to be up-to-date and essentially complete by the end of 2005 for crashes that occurred in 2003 or earlier years. One SCI case from Puerto Rico has been excluded because it has no counterpart in the 50 States + DC FARS.)

TABLE 1-1

RATIO OF SCI CASES[19] TO FARS FRONTAL FATALITIES[20]
(Occupants at seat positions equipped with frontal air bags[21]
in Model Year 1989-1997 vehicles in Calendar Years 1990-2003)

	SCI Fatalities	FARS Frontal Fatalities	SCI Cases per 100 FARS Frontals
Child passengers (age 0-12 years)	146	351	41.60
Adult passengers (age 13+ years)	9	3,899	.23
Drivers	78	29,361	.27

Table 1-1 shows a vast difference between child passengers and adult passengers/drivers. There were 146 SCI cases and 351 FARS frontal fatalities of children age 0-12 years in vehicles with not-yet-redesigned air bags. In other words, close to half (41.60 percent) of the child passenger fatalities are SCI cases. Any remedy that greatly reduces SCI cases should also have a prominent effect in FARS analyses of overall fatality rates. By contrast, only a quarter of one percent of adult fatalities (9 of 3,899 adult passengers and 78 of 29,361 drivers) are SCI cases.[22] A great reduction of SCI cases will be invisible in any statistical analysis of FARS, and can only be detected by directly analyzing the SCI data.

---

[19] Excluding Puerto Rico and Canada.

[20] Initial and/or principal impact 11, 12 or 1:00, excluding first-event rollovers, fires, immersions and similar non-collision events.

[21] All front-seat passengers are included in vehicles with dual air bags – namely, right-front, center-front, sitting on somebody else's lap, other/unknown front-seat position.

[22] An additional factor possibly contributing to this disparity is that NHTSA reviewed every child passenger fatality on FARS and sent SCI teams to investigate any case that appeared likely to have Delta V < 25 mph. No similarly comprehensive review was performed on FARS cases of driver and adult passenger fatalities.

Tables 1-2 and 1-3 examine subgroups of child passengers, identifying who is at the greatest risk. Table 1-2 subdivides children into four age groups. Risk is highest for infants (64 SCI cases per 100 FARS frontals), but it is only slightly lower for toddlers age 1-5 years (51 per 100), and it continues to be grave for 6-10 year-olds (32 per 100). The risk drops greatly at age 11-12 years, to 6 per 100. Nevertheless, that is apparently still much higher than the rate for adult passengers, 0.23 per 100 (however, the rate for 11-12 year-olds is quite uncertain, as it is based on just 2 SCI cases).

TABLE 1-2

CHILD PASSENGERS:
RATIO OF SCI CASES TO FARS FRONTAL FATALITIES, BY AGE GROUP
(Front-seat passengers in vehicles equipped with dual air bags
in Model Year 1989-1997 vehicles in Calendar Years 1990-2003)

Age of the Child Passenger	SCI Fatalities	FARS Frontal Fatalities	SCI Cases per 100 FARS Frontals
Less than 1 year	27	42	64.29
1 – 5 years	80	158	50.63
6 – 10 years	37	117	31.62
11 – 12 years	2	34	5.88

Table 1-3 further subdivides child passengers by age and restraint use. Cases with unknown restraint use (all in FARS) are not included in the calculations. The group with highest risk, as might be expected, is infants in rear-facing safety seats. The number of SCI cases (24) almost equals the number of FARS cases (25). Unrestrained children are at great risk up to age 10. For example, even at age 6-10, there are 47 SCI cases per 100 FARS cases.

The use of a child safety seat or safety belt can substantially reduce risk, but is not a panacea. At age 1-5 and 6-10 years, the risk ratio for restrained children is close to half the ratio for unrestrained. Both SCI cases age 11-12 were belted. The detailed SCI investigations show that many of those child safety seats and safety belts were incorrectly used. For example, the shoulder harness was often routed beneath the arm. Because FARS rarely has similarly detailed investigation of how child safety seats or belts were used, it is impossible to compute separate rates for correctly and incorrectly used systems.

TABLE 1-3

CHILD PASSENGERS: RATIO OF SCI CASES TO FARS FRONTAL FATALITIES
BY AGE GROUP AND RESTRAINT USE
(Front-seat passengers in vehicles equipped with dual air bags
in Model Year 1989-1997 vehicles in Calendar Years 1990-2003)

Age	Restraint Use	SCI Fatalities	FARS Frontal Fatalities	SCI Cases per 100 FARS Frontals
< 1 year	Unrestrained	3	13	23.08
	In a safety seat	24	25	96.00
1 – 5 years	Unrestrained	66	103	64.08
	In a safety seat	6	19	31.58
	Belted (age 2-5)	8	25	32.00
6 – 10 years	Unrestrained	27	58	46.55
	Belted	10	51	19.61
11 – 12 years	Unrestrained	0	12	0
	Belted	2	20	10.00

From Table 1-1, it is obvious that the 9 SCI cases of adult passengers are too few for more detailed analyses, but the 78 drivers are a large enough set to consider subgroups. Any perusal of SCI cases reveals that many of the driver fatalities were females, and that they were often older and shorter than the average driver.[23] By computing ratios of SCI to FARS cases, however, it is possible to quantify the variation in risk among groups of drivers and even compare the highest-risk adults to children. (FARS only began reporting the heights of drivers in calendar year 1998, whereas SCI reports it for all calendar years; thus, we will assume that in calendar years 1990-1997, FARS drivers had the same height distribution as in 1998-2003).

Table 1-4 shows a profound association between a driver's height and the risk of sustaining a fatal injury caused by air bags. Drivers up to 5'3" tall experienced 0.907 SCI cases per 100 FARS fatalities. The risk was half as large, 0.484, for 5'4" – 5'7" drivers, and much lower for drivers ranging from 5'8" up to 6'. There were no SCI cases in pre-1998 vehicles of drivers over 6 feet tall.

---

[23] http://www-nrd nhtsa.dot.gov/departments/nrd-30/ncsa/sci html brings up the latest tables of SCI cases.

8

TABLE 1-4

DRIVERS:
RATIO OF SCI CASES TO FARS FRONTAL FATALITIES, BY DRIVER'S HEIGHT
(MY 1989-1997 vehicles equipped with driver or dual air bags in CY 1990-2003)

Driver's Height	SCI Fatalities	FARS Frontal Fatalities	SCI Cases per 100 FARS Frontals
Up to 5' 3"	35	3,860	.907
5' 4" – 5' 6"	28	5,778	.484
5' 7" – 5' 9"	7	7,449	.094
5'10" – 6'	6	7,879	.076
Over 6'	0	3,442	.0

The effect of gender is difficult to analyze because gender is strongly associated with height. Only two height ranges, 5'4" – 5'6" and 5'7" – 5'9" include substantial numbers of both males and females. Table 1-5 shows twice as much risk for females as males in the 5'4" – 5'6" range; on the other hand, no 5'7" – 5'9" females were SCI cases, but 7 males. Thus, it is not clear if being female adds risk, beyond the fact that females are less tall, on the average, than males.

TABLE 1-5

RATIO OF SCI CASES TO FARS FRONTAL FATALITIES
BY DRIVER'S HEIGHT AND GENDER
(MY 1989-1997 vehicles equipped with driver or dual air bags in CY 1990-2003)

Driver's Height	Gender	SCI Fatalities	FARS Frontal Fatalities	SCI Cases per 100 FARS Frontals
Up to 5' 3"	Male	0	374	.0
	Female	35	3,486	1.004
5' 4" – 5' 6"	Male	6	2,125	.282
	Female	22	3,653	.602
5' 7" – 5' 9"	Male	7	5,668	.124
	Female	0	1,781	.0
5'10" or more	Male	6	10,966	.055
	Female	0	354	.0

Table 1-6 shows that the SCI fatality risk is approximately 3 times as great for drivers age 70 years or older than for drivers 55 years and younger.

TABLE 1-6

RATIO OF SCI CASES TO FARS FRONTAL FATALITIES, BY DRIVER'S AGE
(MY 1989-1997 vehicles equipped with driver or dual air bags in CY 1990-2003)

Driver's Age	SCI Fatalities	FARS Frontal Fatalities	SCI Cases per 100 FARS Frontals
Up to 55 years	38	21,381	.178
56-69 years	14	3,609	.388
70 years and older	26	4,371	.595

Similarly, Table 1-7 indicates that unrestrained drivers have about 1½ times the SCI risk as belted drivers. In other words, safety belts considerably reduce but by no means eliminate the risk of SCI-type fatalities. Furthermore, even correctly used safety belts do not eliminate the risk: according to the detailed SCI investigations, only 5 of the 26 belted fatalities in Table 1-7 had misused the belts. In one sense, Tables 1-6 and 1-7 actually understate the SCI risk for older and unrestrained drivers, by measuring it relative to overall FARS cases. Older and unrestrained drivers also have a higher risk of any type of fatality, given a crash. In more absolute terms – e.g., per 100 towaway crashes – the increase in SCI cases would be proportionately even greater.

TABLE 1-7

RATIO OF SCI CASES TO FARS FRONTAL FATALITIES, BY DRIVER'S BELT USE
(MY 1989-1997 vehicles equipped with driver or dual air bags in CY 1990-2003)

Driver's Belt Use	SCI Fatalities	FARS Frontal Fatalities	SCI Cases per 100 FARS Frontals
Unrestrained	49	14,722	.333
Belted	26	11,908	.218

Finally, Table 1-8 combines all the preceding risk factors and computes the SCI fatality risk of the most vulnerable group of drivers. The average driver has 0.27 SCI cases per 100 FARS frontal fatalities. The risk is almost quadruple, 1.00, for female drivers up to 5'3" tall. It is nearly double that, 1.94, if they are also age 70 years or older. It nearly doubles again, to 3.88 per 100 FARS cases, for unrestrained females, up to 5'3", age 70 or older. That is 14 times the risk for the average driver. Nevertheless, it is still not as high as the risk for 11-12 year-old child passengers (5.88 according to Table 1-2) and it is only a fraction of the risk for children age 6-10

10

years (31.62), let alone toddlers or infants. Furthermore, even in this most vulnerable group of drivers, a great reduction of SCI cases will probably be invisible in any statistical analysis of FARS, and can only be detected by directly analyzing the SCI data.

TABLE 1-8

RATIO OF SCI CASES TO FARS FRONTAL FATALITIES
FOR SELECTED GROUPS OF DRIVERS
(MY 1989-1997 vehicles equipped with driver or dual air bags in CY 1990-2003)

	SCI Fatalities	FARS Frontal Fatalities	SCI Cases per 100 FARS Frontals
All drivers	78	29,361	.27
Female drivers up to 5'3"	35	3,486	1.00
Females ≤ 5'3", age 70+	14	721	1.94
Females ≤ 5'3", age 70+, unrestrained	7	180	3.88

1.4   The 1998-99 redesign of frontal air bags

The design of air bags involves a trade-off between quickly supplying enough gas to absorb a substantial portion of the kinetic energy of a large, unrestrained occupant, and minimizing risk to out-of-position occupants by limiting the force of deployments. In the early 1980s, when belt use was less than 15 percent, the unrestrained occupant was a priority. Air bags had to pass a 30 mph crash test, where the vehicle itself impacts a rigid barrier, with unrestrained dummies in the driver's and right-front seats. Early air bags were located in positions where they could most directly intercept the forward motion of an unrestrained occupant. Furthermore, the trend toward smaller cars after 1973 may have resulted in even more forceful air bags than originally contemplated. Because they have less deformable structure, small cars decelerate more abruptly than large cars in a 30 mph barrier test, and the air bag must fully deploy and be ready to cushion the occupant even sooner.

By 1995-1996, serious injuries to children and other out-of-position occupants urgently required mitigation. In the meantime, use of safety belts had increased to 61 percent and it had become the law in 49 States and DC.[24]

At an April 12, 1995 SAE meeting, and in subsequent briefings to NHTSA, Dr. Prasad of Ford presented a rationale for redesigning air bags:

- "Aggressive" air bags are injuring out-of-position occupants and even some correctly positioned occupants.

---

[24] Kahane (2004), pp. 90-91; *Observed Safety Belt Use in 1996*, NHTSA Research Note, Washington, 1997; *Third Report to Congress – Effectiveness of Occupant Protection Systems and Their Use*, NHTSA Report No. DOT HS 808 537, Washington, 1996.

- Reducing the initial velocity of the deployment and other technical changes can significantly lower the injury risk.

- Ford believes compliance with the unrestrained 30 mph crash test required by FMVSS No. 208 would be impossible or quite difficult with the foreseeable technologies to make air bags less aggressive.

- NHTSA should ease the unrestrained test in some way to allow the redesign. He recommended reducing the speed to 25 mph.

- The great increase in belt use, with buckle-up laws in almost every State justifies shifting the highest priority from the unrestrained to the belted occupant.

Specifically, the initial velocity of deployments can be reduced by "**depowering**": by removing some of the gas-generating propellant or stored gas from the inflator. There is no consensus parameter for quantifying the extent of depowering. Two parameters cited in the literature are the percent reduction in the peak pressure of the inflator and the reduction in the average rise rate of the pressure as it approaches its peak.

NHTSA redirected its ongoing computer simulation of air bag deployments to study possible effects of depowering. On November 9, 1995, the agency issued a request for comments and announced a public meeting to discuss methods to reduce adverse effects of air bags, including depowering. At that time, with Ford's permission, NHTSA docketed Dr. Prasad's briefing notes; the agency also presented a response, plus the results of its own computer simulations.[25]

NHTSA developed a program of 89 sled tests and other laboratory tests to systematically evaluate the safety performance of depowered air bags. The tests were conducted during April-August 1996.[26] The findings are summarized later in this section. They demonstrated that depowered air bags were a satisfactory interim measure that could be implemented quickly to reduce the adverse effects of air bags.

On August 6, 1996, while the test program was still underway, the agency issued a Notice of Proposed Rulemaking (NPRM) pledging to revise FMVSS No. 208 to reduce adverse effects of air bags. One of the possible revisions would be to ease the unrestrained test in order to allow depowering and possibly other redesign. Public response to the earlier 11-9-1995 notice raised doubts whether a reduction of the test speed from 30 to 25 mph, as proposed by Ford, would allow sufficient reductions in the initial velocity of air bags. NHTSA suggested, instead, leaving the test speed at 30 mph but allowing 80 rather than 60 chest g's on the dummies.[27]

On August 23, 1996, the American Automobile Manufacturers Association (AAMA) sent NHTSA a Petition for Rulemaking to replace the 30 mph unbelted crash test in FMVSS No. 208 with a sled test, also at 30 mph, using a standardized crash pulse over a duration of 143 milliseconds. On November 13, 1996, AAMA revised its petition, proposing a more severe crash pulse with duration of 125 milliseconds and a peak sled acceleration of approximately 17.2

---

[25] *Federal Register* 60 (November 9, 1995): 56554; *Status Report: On the Issue of Testing Air Bag-Equipped Vehicles with and without Belt Restraints at Different Speeds*, NHTSA Docket No. NHTSA-1996-1772-2, 1995.
[26] http://www-nrd.nhtsa.dot.gov/departments/nrd-11/airbags/abgdb/
[27] *Federal Register* 61 (August 6, 1996): 40784.

g's, and also requesting measurement of neck injury criteria. This crash pulse resembles the performance of large cars of the 1970s in barrier impacts – i.e., more gradual than the typical car of the 1990s. AAMA argued that their proposal would:

- Allow a more extensive depowering or redesign of air bags than the alternatives of lowering test speed or raising chest g's on a full-vehicle crash test,

- Maintain an adequate level of protection for the unrestrained occupant, and

- Expedite the redesign of air bags to reduce their adverse consequence, because it is simpler and quicker to design and control performance on a sled test than on a crash test.

NHTSA agreed that the AAMA proposal would expedite the redesign of air bags (while reserving judgment whether depowering could have been accomplished, over a longer time span, without changing the unrestrained test). On January 6, 1997, the agency issued an NPRM incorporating the sled test in AAMA's revised petition.[28] The NPRM was widely approved by the safety community. On March 19, 1997, NHTSA issued its Final Rule on Depowering. It amended FMVSS No. 208, effective immediately. Until at least September 1, 2001 (later extended to September 1, 2006), manufacturers would have a choice: they could now certify a vehicle complied with the unbelted 30 mph sled test with standardized crash pulse, or they could certify, as before, that the vehicle complied with the unbelted 30 mph crash test.[29] A make-model is called "**sled-certified**" if the manufacturer has chosen the sled test option. The manufacturers quickly exercised that option. Make-models comprising approximately 70 percent of sales were sled-certified for the entire 1998 model year. By the beginning of the 1999 model year, about 99 percent were sled-certified. The regulation did not affect the 1997 model year because its certification and production set-up had been completed before March 19, 1997; similarly, if 1998 models were not sled-certified for the full model year, it was usually because their production cycle was already well underway by March 19, 1997.

Test program findings NHTSA's Final Regulatory Evaluation of Depowering summarizes the results of 89 deployment tests conducted in April-August 1996 and draws inferences on the potential effects of depowering on injuries in crashes.[30] The agency acquired multiple copies of air bag systems in selected production vehicles and installed them in sled bucks reproducing those vehicles' interiors, with dummy occupants. Driver or passenger air bags were deployed:

- In their original condition, or depowered from 18 to 60 percent (as measured by the reduction in the peak pressure of the inflator) by removing some of the propellant.

- With the passenger air bag mid-mounted or top-mounted in the instrument panel, depending on how it was originally installed in the vehicle.

---

[28] *Federal Register* 62 (January 6, 1997): 807.
[29] *Federal Register* 62 (March 19, 1997): 12960; *Code of Federal Regulations*, Title 49, Part 571.208 S13. The sled test option expired on September 1, 2006 because at that time all new vehicles had advanced air bags. The unrestrained test for advanced air bags is again a barrier test, not a sled test; however, the barrier test is now performed at 25 mph with 5th percentile female or 50th percentile male dummies, unlike the barrier test for original air bags, performed at 30 mph, but only with 50th percentile male dummies.
[30] *Final Regulatory Evaluation, Depowering*, NHTSA Docket No. NHTSA-1997-2817-002, 1997.

- With dummies ranging from 12 month-old, 3 year-old, 6 year-old, 5th percentile female to 50th percentile male.

- With correctly positioned adult dummies – belted and unrestrained – the location of the seat ranging from all the way back to all the way forward.

- With infant dummies in rear-facing safety seats. However, tests were performed only with air bags in their original condition, not depowered.

- With out-of-position child dummies standing directly in front of the instrument panel, leaning over it, or sitting on the floor directly in front of it. Fifth-percentile female dummies sat directly in front of the steering wheel or leaned over it.

Feasibility of depowering: NHTSA agreed with AAMA that shifting from the unrestrained 30 mph crash test to the generic sled test would allow an immediate depowering in the range of 20-35 percent in essentially all make-models. More extensive depowering might result in failures (on chest g's) on the sled test, and were not considered further in the analysis.

Infants in rear-facing safety seats: The tests with the 12 month-old dummy did not compare the performance of depowered and original air bags. NHTSA was also reluctant to draw inferences about infants from the depowered tests with the 3 year-old dummy. The agency did not estimate any specific benefit from depowering for infants in rear-facing seats. As stated above, infants should never ride in the front seat at a position equipped with a working air bag.

Out-of-position child passengers: In actual crashes, out-of-position children had high risk of neck and head injuries. The deployment tests showed high neck injury criteria for the dummies. Depowered air bags resulted in substantially lower neck injury criteria. NHTSA translated neck injury criteria into probabilities of fatal injury and averaged the results for the various sled bucks and dummy positions. The agency expected that 20-35 percent depowering of air bags could reduce fatalities to out-of-position children by 43.5 percent.

Out-of-position drivers and adult passengers: Fatalities in actual crashes involved head, neck and/or chest injuries. NHTSA conducted out-of-position tests with 5th percentile female dummies in the driver's seat. They included tests specifically comparing 25-42 percent depowered air bags to original air bags in two make-models and other tests comparing 1994 and 1996 vehicles of the same make-model where air bags had been redesigned. The tests generally showed reductions in head, neck and chest injury criteria with depowered and redesigned air bags. Formulas were not available to translate some of these criteria into probabilities of fatal injury. Thus, NHTSA did not quantify a percentage reduction in fatalities but anticipated that depowering "could reduce a large portion, but not all of these fatalities" on the driver's side and "could eliminate almost all of these fatalities" among adult passengers.[31]

Correctly positioned adult drivers and passengers: In directly frontal impacts that closely resemble a collision with a barrier – resulting in abrupt deceleration but little intrusion on the vehicle – adult drivers and passengers have a substantial risk of chest injury. Substantial gas pressure is needed in the fully deployed bag to cushion an occupant's thorax, especially if the occupant is unbelted. Depowering, intuitively, could increase chest injury for correctly

---

[31] *Ibid.,* Executive Summary.

positioned, especially unrestrained adults: they might be more likely to "punch through" a lower-pressure air bag and engage structures beneath it, such as the steering assembly. Indeed, sled and crash tests showed increased chest g's with 25 percent depowering, except for belted passengers, where they decreased a little. NHTSA developed two statistical methods to predict the percentage increase in fatality rates per incremental chest g, and two hypotheses about what portion of fatal frontal collisions "closely" resemble a barrier impact. That produced a range of likely effects of depowering:

- Unbelted drivers: 0.04 – 0.6 percent increase in fatalities
- Belted drivers: 0.3 – 2.7 percent increase
- Unbelted passengers: 1.9 – 15.6 percent increase
- Belted passengers: 0.3 – 1.6 percent reduction
- All of the above: 0.3 – 2.8 percent increase

In other words, the overall increase in fatalities to correctly positioned adults might be so small (0.3 percent) as to escape detection in any statistical evaluation of crash data, or it might be large enough (2.8 percent) for possible detection after many years of crash data became available. If the increase were near the lower end of the range, NHTSA estimated that it would be lower, in absolute terms, than the number of children saved by depowering. Even at the high end, a 2.8 percent increase would be small compared to the 29 percent effectiveness of air bags in directly frontal impacts.[32]

In addition to the preceding analyses, the regulatory evaluation cited crash experience of the Holden Commodore in Australia. The original air bags on that make-model had performance characteristics similar to what was being proposed as depowered air bags in the United States. Based on limited data, serious injury rates for belted occupants with air bags were low compared to belted occupants in the same make-model, not yet equipped with air bags. That could indicate depowering might be more beneficial for belted drivers and passengers than was indicated by the analysis of the NHTSA tests.

How the manufacturers responded As stated above, approximately 70 percent of new vehicles were sled-certified for the entire 1998 model year, and 99 percent by model year 1999. But not all the vehicles were depowered or otherwise redesigned at that time. In general, FMVSS are performance standards. Vehicles must pass a test that demonstrates a certain level of safety. If, for example, a vehicle could already pass the test before the FMVSS went into effect, the manufacturer is not obligated to change anything. On the other hand, vehicles may at any time (not just upon the effective date) be improved to a level of safety beyond the test requirement, even if they already met the basic requirement. Sled certification is yet a little more complicated than the usual FMVSS because:

- Manufacturers had the choice of sled-certifying or continuing to use the crash test.

- The sled test offered practical advantages over the crash test. Furthermore, most or all vehicles meeting the crash test could also meet the sled test. Therefore, make-models

---

[32] Kahane (2004), p. 110.

were sled-certified even if their air bags were not aggressive before 1998 and not redesigned in 1998.

In December 1997, NHTSA sent an Information Request (IR) asking manufacturers to describe the performance of driver and passenger air bags in each make-model, year-by-year from the initial installation of air bags up through 1998. Requested parameters included the peak pressure of the deploying air bag; the average rise rate of the pressure as it approaches its peak; volume and rearward extent of the deploying air bag; the threshold speed (target minimum Delta V to trigger deployment); the inflator technology; the location before deployment; and the arrangement of tethers. Many of the individual data were provided by the manufacturers as proprietary, confidential information. However, NHTSA computed fleet-wide averages of selected parameters. The agency's report on *Air Bag Technology in Light Passenger Vehicles* tracked these averages year-by-year and demonstrated how the industry responded in model year 1998.[33] Some of its findings have been updated here by using sales-weighted rather than arithmetic averages across make-models. Peak pressure and rise rate has been "scaled" – i.e., adjusted for air bag volume.

Peak pressure and rise rate – driver air bags: Among the 14 million vehicles sold in model year 1998, nearly 13 million were in make-models that sled-certified at the beginning or middle of model year 1998. The IR's furnished detailed data about air bag performance for nearly 11 million of these 13 million vehicles. They showed that:

- 84 percent were depowered upon sled certification, as evidenced by a reduction in the peak pressure and/or the rise rate.

- 1½ percent were not depowered, but were substantially redesigned as evidenced by a change in another key parameter.

- 7 percent increased peak pressure and rise rate from 1997 (usually by a small amount).

- 7½ percent were unchanged from 1997; however, nearly half of these had been depowered in 1996 or 1997 while still certifying to the crash test.

For all 11 million vehicles with IR data (including those that were not depowered), the peak pressure of driver air bags was reduced by 13 percent, from a sales-weighted fleet-wide average of 186.8 kPa in model year 1997 to 162.5 kPa in 1998. The rise rate was reduced by 24 percent: from 7.46 kPa/msec in 1997 to 5.64 kPa/msec in 1998. For the 84 percent of these vehicles that specifically depowered when they sled-certified, peak pressure dropped by an average of 16 percent, and the rise rate by 30 percent. In other words, for the make-models that actually depowered when they sled-certified, the 16 percent average reduction of peak pressure is slightly less than the 20-35 percent predicted by NHTSA and AAMA, but the 30 percent reduction of the rise rate is in that range.

---

[33] Hinch, J., Hollowell, W.T., Kanianthra, J., Evans, W.D., Klein, T., Longthorne, A., Ratchford, S., Morris, J., and Subramanian, R., *Air Bag Technology in Light Passenger Vehicles,* National Highway Traffic Safety Administration, Washington, 2001, access from http://www-nrd.nhtsa.dot.gov/departments/nrd-11/NRDpubs.html .

Peak pressure and rise rate – passenger air bags: Over 13 million of the 14 million vehicles sold in model year 1998 were in make-models sled-certified at the beginning or middle of model year 1998. IR data were available for 10 million of these 13 million vehicles. They showed that:

- 69½ percent were depowered upon sled certification.

- 11 percent were not depowered, but were substantially redesigned in some other way, most often by shifting from pyrotechnic to hybrid inflators.

- Only ½ percent increased peak pressure and rise rate from 1997.

- 19 percent were unchanged from 1997; most of these had not changed in preceding years, either.

For all 10 million vehicles with IR data (including those that were not depowered), peak pressure of driver air bags was reduced by 13 percent, from a sales-weighted fleet-wide average of 328.6 kPa in model year 1997 to 285.4 kPa in 1998. The rise rate was reduced by 18 percent: from 8.63 kPa/msec in 1997 to 7.09 kPa/msec in 1998. For the 69½ percent of these vehicles that specifically depowered when they sled-certified, peak pressure dropped by an average of 18 percent, and the rise rate by 27 percent. That is fairly consistent with the 20-35 percent predicted by NHTSA and AAMA.

Other parameters – driver air bags: Throughout the 1990s, driver air bags became more often recessed beneath the steering wheel and less often protruding beyond it. Volume of the air bag stayed about the same. Rearward extent of the deploying air bag decreased from an average 16 to 14 inches. Almost all air bags were pyrotechnic and used Sodium Azide. In the early 1990s, the majority of air bags had no tethers; by 1998, 88 percent had two or more tethers. Recessing the bag, reducing its rearward extent, changing the fold pattern and adding tethers would help keep a driver away from the deploying air bag; however, changes were gradual and, unlike depowering, not concentrated in 1998.

Other parameters – passenger air bags: From 1993 to 1998, air bags were modified not to come nearly as close to passengers; the average distance from the furthest rearward extent of deploying mid-mounted air bags to the Seating Reference Point (SRP) increased by 9 inches. Air bag volume gradually decreased by 26 percent from 1993 to 1998 and mass by 10 percent. Air bags were originally pyrotechnic; by 1998, about half of them used a hybrid inflator combining stored gas and a pyrotechnic agent. These changes ought to help reduce injuries from air bags; however, they were gradual and not concentrated in 1998.[34]

Test results: NHTSA followed up the test program of 1996 by performing comparable tests on selected model year 1998 and 1999 vehicles and comparing the injury criteria on the dummies to the previous results for the 1996 vehicles. These tests are not part of the certification procedure for FMVSS No. 208; thus, a score exceeding the IARV (injury assessment reference value for FMVSS No. 208) does not indicate a failure to comply.

---

[34] *Ibid.,* pp. 5-12 and Appendix A.

For out-of-position 5th percentile female driver dummies, the neck injury criteria were substantially lower in 1998-1999 than in 1996; however, they still exceeded the IARV on some of the tests. (Head and chest injury criteria were low in all years.)

Likewise, for out-of-position 6-year-old child passenger dummies, head, neck and chest injury criteria were all substantially lower in 1998-1999 than in 1996; nevertheless they still exceeded the IARV on some tests.

NHTSA also performed 30 mph barrier impacts with unbelted, correctly positioned, 50th percentile male dummies in the driver and right-front seats. The configuration was similar to FMVSS No. 208 compliance tests for pre-1998 vehicles, but these tests were not required for compliance in 1998-1999, because the vehicles had been sled-certified. HIC and chest deflection did not exceed the IARV on any of the 13 drivers and 13 passengers. Chest g's exceeded the IARV on one passenger and femur load, on one driver. Compared to average performance on compliance tests of pre-1998 vehicles equipped with air bags, chest g's had increased on the order of 5-10 percent for drivers and passengers, whereas chest deflection and HIC stayed about the same. In other words, the follow-up tests were consistent with earlier predictions that depowering would substantially reduce injury criteria for out-of-position children and adults while somewhat increasing chest g's for unbelted, correctly positioned adults.[35]

1.5   Other and subsequent actions to reduce risk from air bags

Moving children to the back seat: The most immediate action to reduce risk to children from air bags was launched in October 1995. NHTSA, the manufacturers and the safety community joined in a continuing campaign to inform the public about the hazard of air bags to children and the need for children to ride in the back seat of vehicles with air bags.[36]

The public has responded, at least for the younger children. NHTSA analyzed the seat positions of 363,579 child passengers in crash data from Florida, Maryland and Utah. In 2001, only 8 percent of 0-3 year-old infants and toddlers still rode in the front seat, down from 26 percent in 1995. The proportion of 4-7 year-old children in the front seat had decreased to 19 percent from 33 percent. However, the proportion of 8-12 year-old children in the front seat only declined to 35 percent from 39 percent. In other words, based simply on the reduction in exposure – even before taking the effect of redesigned of air bags into account – we might expect fatalities from air bags to decrease from 1995 to 2001 by about two-thirds for infants and toddlers age 0-3 years, by half for children age 4-7 years, but only a little for pre-teens age 8-12 years.[37]

On-off switches: Since June 22, 1995, FMVSS No. 208 has permitted on-off switches for the passenger air bag in vehicles without back seats or that cannot accommodate rear-facing child

---

[35] *Ibid.,* pp. 26-36 and Appendix D.

[36] *Safety Agency Issues Warning on Air Bag Danger to Children* (1995); *Air Bag & Seat Belt Safety Campaign, Air Bag & Seat Belt Safety Tips*, National Safety Council, Chicago, November 17-30, 2003, http://www.nsc.org/partners/safetips.htm; *Buckle-Up America Child Passenger Safety Week, February 10-16, 2002, Talking Points*, NHTSA, http://www.nhtsa.dot.gov/people/injury/airbags/buckleplan/CPS%20Week%20Planner_files/talking1.html.

[37] Kahane (2004), pp. 134-135; Kindelberger, J. and Starnes, M., *Moving Children from the Front Seat to the Back Seat: The Influence of Child Safety Campaigns*, NHTSA Research Note No. DOT HS 809 698, Washington, 2003.

safety seats in the back seat.[38] In such vehicles, on-off switches will remain an option until model year 2012. Almost all pickup trucks through model year 2004 except full crew cabs have on-off switches for their passenger air bags – and the air bags were mostly introduced in 1997 or 1998. On-off switches are located on the instrument panel and can only be operated by the ignition key. Switches have two settings, "air bag on" and "air bag off."[39] When the switch is "off," a light on the dashboard advises "air bag off," and that light stays on as long as the ignition is on.

For switches to be fully effective, they need to be turned "off" whenever the right-front passenger is a child age 0-12 years and "on" if the passenger is 13 years or older (with rare exceptions). NHTSA's 2000 survey of pickup trucks in operation showed that on-off switches to a large extent prevented the exposure of small children to air bags while preserving the life-saving benefits of air bags for teenagers and adults. Nevertheless, public use of the switches was not nearly 100 percent as recommended. In particular, switches were left on for a large proportion of child passengers age 7-12 years, but turned off for nearly half of adults 70 years or older (the percentage of switches at the recommended setting is shown in bold type):[40]

Age of the Passenger	Switch "On" (%)	Switch "Off" (%)
Less than 1 year[41]	14	**86**
1-6 years	26	**74**
7-8	41	**59**
9-10	53	**47**
11-12	70	**30**
13-15	**78**	22
16-19	**83**	17
20-59	**85**	15
60-79	**81**	19
70 years and older	**44**	56

Keeping drivers away from air bags: A public information campaign advised drivers to sit at least 10 inches away from the air bag while operating the vehicle. In some vehicles, pedal adjusters or other modifications made it easier for short drivers to keep a distance from the air bag. Belt use increased from 61 percent in 1996 to 75 percent in 2002, helping drivers stay back

---

[38] *Federal Register* 60 (May 23, 1995): 27233; *Code of Federal Regulations*, Title 49, Government Printing Office, Washington, 2005, Part 571.208 S4.5.4. See also: *NHTSA Permits Air Bag Switch to Prevent Injury to Infants in Rear-Facing Safety Seats*, Press Release No. NHTSA 30-95, U. S. Department of Transportation, Office of the Assistant Secretary for Public Affairs, Washington, 1995; the press release, dated May 18, 1995, still says "air bags offer excellent supplemental protection to … most children" but "pose a unique safety risk to infants in rear-facing child seats."

[39] In model year 2003, General Motors introduced switches with a third setting, "auto" for advanced air bags.

[40] Kahane (2004), pp. 115-117; Morgan, C., *Results of the Survey on the Use of Passenger Air Bag On-Off Switches*, NHTSA Technical Report No. DOT HS 809 689, Washington, 2003, pp v and 4-5.

[41] 29 right front passengers less than 1 year old and/or seated in a rear-facing child safety seat, 4 switches "on." In these four cases, the driver was not the owner of the truck, or mistakenly believed the switch was "off." Not a single driver intentionally left the switch "on" in his or her own truck.

in their seats before the principal impact. Starting on January 19, 1998, high-risk individuals could obtain aftermarket on-off switches for the driver air bag at their own expense, but as of August 2002, only 12,513 vehicles were known to have any type of aftermarket switches.[42]

Dual-stage air bags: In 1999, a few make-models were equipped with dual-stage air bags on the passenger side. By model year 2001, close to 20 percent of new vehicles were so equipped for drivers and passengers. Dual-stage inflators tailor the amount of pressure in the frontal air bag during a crash. In severe crashes, both stages go off at the same time, resulting in a higher-pressure air bag deployment for maximum absorption of the occupant's kinetic energy. But in less severe crashes requiring less inflation force to cushion the occupant, only one stage of the inflator may go off. That results in a lower-pressure air bag deployment and reduces risk to out-of-position occupants. Some vehicles take into account the occupant's belt use, weight and seat position as well as the anticipated severity of the crash. Generally, if the occupant is belted, it takes a higher crash severity to trigger both stages of the inflator, because less cushioning force is needed than for an unbelted occupant.

Advanced air bags: In 2000, NHTSA amended FMVSS No. 208 to make future air bags substantially less hazardous to out-of-position occupants, but also more effective for correctly positioned occupants. These "advanced" air bags are being implemented step by step. From September 1, 2003 to September 1, 2006, air bags were phased in that do not deploy at all for children ("suppression"), deploy only at a low level of force ("low-risk deployment"), or track an occupant's motion and suppress the air bag if they are too close ("dynamic automatic suppression"). The first-generation technology for suppression and low-risk deployment consists of sensors that essentially discriminate between child and adult passengers based on their weight and/or size while they are seated. The sensors may simply measure the weight of the occupant, or they may be embedded in pattern-recognition pressure pads located below the seat cushion. The sensors determine by the intensity and distribution of the pressure around the cushion whether the seat is occupied by a child in a safety seat, a child not in a safety seat, or an adult. Future systems may detect an occupant's position and ascertain if anybody is close to the air bag in the moment before impact. (As of August 2006, none use such technology.) Furthermore, advanced air bags need to pass a barrier and offset test with 5th percentile female dummies in addition to the original barrier test with a 50th percentile male dummy. The unrestrained 30 mph sled test was superseded by a barrier test, but at a speed of 20-25 mph. MY 2008-2010 will phase in a 35 mph barrier test with the belted 50th percentile male dummy, an increase from the current 30 mph.[43]

Pretensioners for safety belts: By model year 2002, approximately 63 percent of new vehicles were equipped with pretensioners that retract the safety belt almost instantly in a crash to remove excess slack. They are mechanical or pyrotechnic devices located within the belt's retractor or buckle assemblies. By pulling in slack, they can reduce belted occupants' potential for

---

[42] *Federal Register* 62 (November 21, 1997): 62406; *Air Bags & On-Off Switches*; Kahane (2004), pp. 90 and 115.
[43] *Federal Register* 65 (May 12, 2000): 30679; *Code of Federal Regulations*, Title 49, Part 571.208 S14. NHTSA has also proposed to phase in, on the same schedule, a 35 mph test with 5th percentile female dummies – *Federal Register* 68 (August 6, 2003): 46539.

interacting with deploying air bags. Of course, they would have no effect on unbelted occupants.[44]

1.6   Earlier statistical analyses of redesigned air bags

Based on engineering analyses and test results, NHTSA and the safety community's expectations about redesigned air bags included:

- Optimism that fatalities and injuries to out-of-position occupants would decrease substantially except, perhaps, to infants in rear-facing seats. A statistically significant reduction in the rate of these fatalities would be needed to demonstrate that the redesign accomplished its goal.

- Concern that fatalities to correctly positioned but unbelted adults could increase in frontal crashes, relative to pre-1998 air bags. Here, if statistical analyses based on adequate data fail to show a significant change, we could still say that redesign accomplishes its goal of preserving the life-saving benefits of air bags. No news would be good news.

- Uncertainty about the effect on correctly positioned, belted occupants: it might be beneficial, harmful, or most likely, negligible. Here, too, no news would be good news.

Analyses of child and/or out-of-position fatalities *NHTSA SCI tabulations*: The Special Crash Investigation (SCI) program identifies cases of fatalities due to occupant contact with air bags in otherwise survivable crashes (Delta V < 25 mph). Every quarter, NHTSA tabulates fatalities to date by vehicle model year and computes the fatality rate, by model year, per million cumulative vehicle registration years.[45] The tables show dramatic reductions throughout the later 1990s. For example, as of January 1, 2006, the SCI fatality rate for child passengers was 0.366 in model year 1996, 0.281 in 1997, 0.074 in 1998 and 0.038 in 1999. In other words, for children, the reduction is especially large in 1998. The SCI fatality rate for adult drivers had decreased from 0.108 in 1996 to 0.020 in 2000, but did not show a particular reduction in 1998. The basic trend has already been obvious for years. In other words, the SCI tabulations demonstrate that something is greatly reducing the fatalities to out-of-position occupants, but it is uncertain how much of the effect is due to redesign vs. other factors such as public information campaigns.

NHTSA analyses of SCI data: In its 2000 economic assessment of advanced air bags, the agency analyzed the SCI data available at that time, comparing fatality rates by calendar year as well as model year, in an attempt to sort out the effects of redesigned air bags and public information campaigns.[46] Overall, the SCI fatality rate was 65 percent lower for model year 1998 than for 1996-1997. But when the data were limited to the same calendar year period (10/01/1997 – 1/01/2000), the fatality rate was just 44 percent lower in model year 1998 than in 1996-1997. The agency concluded that "about 2/3 of the [overall] benefit is from redesign of the air bags and 1/3 of the benefit is from changes in behavior" such as moving children to the back seat or

---

[44] Kahane (2004), pp. 103-104; Walz, M.C., *NCAP Test Improvements with Pretensioners and Load Limiters*, NHTSA Technical Report No. DOT HS 809 562, Washington, 2003.

[45] NHTSA National Center for Statistics and Analysis, Special Crash Investigations, Normalized Tables and Charts; access from http://www-nrd.nhtsa.dot.gov/departments/nrd-30/ncsa/sci html .

[46] *Final Economic Assessment, FMVSS 208, Advanced Air Bags*, NHTSA Docket No. NHTSA-2000-7013-2, 2000, p. II-9.

drivers sitting further from the steering wheel. In 2003, Kindelberger, Chidester and Ferguson analyzed the latest SCI data and again concluded that the redesign of air bags and changes in behavior both made substantial contributions to reducing fatalities.[47] They compared SCI fatality rates with barrier-certified and sled-certified air bags, by calendar year. The analyses in Chapter 2 of this report use a similar approach with more recent SCI data.

Analysis of insurance data: Researchers at the Children's Hospital of Philadelphia and at State Farm Insurance collected and analyzed information on 1998-2002 frontal crashes involving a restrained child passenger age 3-15 years in the front seat.[48] They compared, for redesigned vs. original air bags, the number of deployments per 100 frontal crash involvements and the odds of moderate (AIS ≥ 2) injury for children involved in crashes where the air bag deployed. In passenger cars, the injury odds ratio was a statistically significant 51 percent lower with redesigned air bags than with original air bags; in minivans, the reduction was 52 percent (not significant). However, in SUVs, the injury odds ratio was 30 percent higher with redesigned than with original air bags (also not significant).

The inclusion of injuries that are not life-threatening (AIS 2,3) and 13-15 year-old teens, and the exclusion of unrestrained children as well as all 0-2 year-old infants and toddlers removes the analysis somewhat from the issue of serious injuries to out-of-position, mostly unrestrained children. The computation of injury rates per 100 *deployments* is an additional source of potential bias, because the probability of deployment can vary from model to model and can change as models are redesigned. The large and similar injury reductions in passenger cars and minivans are encouraging, but the results for SUVs raise a flag and will require additional analysis in this report.

Analyses of overall effectiveness for adults *1998-1999 FARS analysis*: One technique for estimating the overall effectiveness of air bags is to compare the ratio of frontal to non-frontal fatalities in vehicles equipped with air bags to the corresponding ratio in vehicles without air bags; the non-frontal fatalities act as a control group.[49] The same technique can be used to estimate the change in fatality risk from one type of air bag to another – e.g., redesigned vs. original. In 2000, as part of its economic assessment of advanced air bags, NHTSA compared fatality risk of front-outboard occupants based on FARS data for 1998 and the first six months of 1999:[50]

---

[47] Kindelberger, J.C., Chidester, A.B., and Ferguson, E., "Air Bag Crash Investigations," *Proceedings 18th International Technical Conference on the Enhanced Safety of Vehicles*, NHTSA Report No. DOT HS 809 543, Washington, 2003, Paper No. 299.
[48] Arbogast, K.B., Durbin, D.R., Kallan, M.J., and Winston, F.K., "Effect of Vehicle Type on the Performance of Second Generation Air Bags for Child Occupants," *47th Annual Proceeding – Association for the Advancement of Automotive Medicine*, Barrington, IL, 2003, pp. 85-99.
[49] Kahane (2004), pp. 109-110; *Fifth/Sixth Report to Congress – Effectiveness of Occupant Protection Systems and Their Use*, NHTSA Report No. DOT HS 809 442, Washington, 2001, pp. 6-11.
[50] *Final Economic Assessment, FMVSS 208, Advanced Air Bags*, p. II-13.

	Frontal Fatalities	Non-Frontal Fatalities	Frontal/Non-Frontal Risk Ratio
Original air bags (MY 1995-1997)	3,684	2,699	1.365
Redesigned air bags (MY 1998-2000)	1,051	782	1.344

The ratio of frontal to non-frontal fatalities stayed almost the same; in fact, it had improved by a non-significant 2 percent in the vehicles with redesigned air bags. This was the first, important bulletin of "no news that was good news," that redesigned air bags were preserving the life-saving benefits of air bags for correctly positioned adults. The analyses of this report will update that approach and allow considerable refinement because so many more data have become available.

2000-2002 FARS analysis: Braver, Kyrychenko and Ferguson compared frontal fatality rates per million registration years for vehicles with redesigned vs. original air bags, of model years 1997-1999.[51] Fatality counts were obtained from 2000-2002 FARS. R.L. Polk supplied registration data, supplemented by information from the National Household Travel Survey and insurance files to adjust for annual mileage or subdivide by gender. "Frontal" crashes are defined as those having 12:00 impact point on FARS. Drivers' fatality rates were 6 percent lower with redesigned air bags than original air bags. Whereas this overall reduction is not statistically significant, it is additional, reassuring evidence that the redesign of air bags is not depriving adults of overall crash protection. One finding that definitely needs a follow-up with additional data is that fatality rates in pickup trucks, unlike the other vehicle types, significantly increased with redesigned air bags. Additional data would also make it possible to focus the analysis more closely on redesigned air bags, for example, by removing make-models whose air bags remained unchanged upon sled certification, or that changed other features such as safety belt pretensioners.

Other statistical analyses: A Blue Ribbon Panel for Evaluation of Depowered and Advanced Air Bags, drawn from the safety research community, presented findings based on smaller datasets than FARS at a public meeting and a technical conference in 2003.[52] In-depth crash investigations by the University of Michigan "suggest that 1998 and later model-year vehicles are at least as effective as pre-1998 model-year vehicles in protecting the head, neck, chest, face, and abdomen of belted and unbelted occupants in moderate-to-severe frontal crashes."[53] Analyses of NHTSA's Crashworthiness Data System for 1993-2001 indicated that "drivers in frontal crashes of known severity in 1998-2002 model year vehicles sustained significantly fewer and less severe injuries than their counterparts in pre-1998 model year vehicles. This was true of all drivers, males, females, MAIS 2+ and 3+, and across crash severity."[54] However, in crashes investigated by the Ryder Trauma Center in Florida, "preliminary data show an 8 percent higher

---

[51] Braver, E.R., Kyrychenko, S.Y., and Ferguson, S.A., "Driver Mortality in Frontal Crashes: Comparison of Newer and Older Air Bag Designs," *Traffic Injury Prevention*, Vol. 6, March 2005, pp. 24-30.

[52] Ferguson, S.A., Schneider, L., Segui-Gomez, M., Arbogast, K., Augenstein, J., and Digges, K.H., "The Blue Ribbon Panel on Depowered and Advanced Airbags – Status Report on Airbag Performance," *47th Annual Proceeding – Association for the Advancement of Automotive Medicine*, Barrington, IL, 2003, pp. 79-102.

[53] *Ibid.*, p. 83.

[54] *Ibid.*, p. 84.

overall fatality rate with later model airbags."[55] Summarizing these findings as well as the FARS and SCI analyses completed by 2003, Dr. Ferguson, the panel's moderator, concluded that "There has been no large reduction in the [overall] effectiveness of airbags as some had predicted. Indeed, most of the statistical analyses that have been conducted to date indicate there has been a small but measurable increase in [overall] effectiveness.... Depowered ... air bag systems have dramatically reduced the harm to out-of-position children and adults in low-speed crashes."[56]

---

[55] *Ibid.*, p. 100.
[56] *Ibid.*, p. 101.

CHAPTER 2

EFFECT OF REDESIGNED AIR BAGS ON SCI FATALITY RATES

2.0   Summary

NHTSA's Special Crash Investigation (SCI) program identifies cases of fatalities due to occupant contact with air bags, as evidenced by detailed analysis of medical records and the vehicle interior, in otherwise survivable crashes, as evidenced by a Delta V estimated to be less than 25 mph. The redesign of air bags in 1998-1999 was highly effective in reducing these low-to-moderate speed fatalities. The SCI fatality rate per billion vehicle registration years in calendar years 1998-2003 for 0-12 year-old child passengers is 83 percent lower in vehicles with redesigned air bags than in vehicles with original air bags. The reduction for drivers is 70 percent, and for adult passengers, 42 percent.

Even greater is the combined effect of redesigned air bags and public information campaigns urging that children sit in the back seat and adults sit as far as possible from air bags. The SCI fatality rate for child passengers with redesigned air bags in calendar years 1998-2003 is 93 percent lower than the rate with original air bags in calendar years 1990-1997 (i.e., before the campaigns really took effect). The corresponding reduction for drivers ranges up to 95 percent, and for adult passengers, 90 percent.

2.1   SCI fatality rates for child passengers

By the end of calendar year 2005, NHTSA's Special Crash Investigations included records of 166 children age 0-12 years with fatal injuries from air bags in otherwise survivable, low-to-moderate speed crashes (Delta V < 25 mph): 24 infants in rear-facing child safety seats and 142 children not in rear-facing child safety seats. In each of these crashes, the make-model and model year of the vehicle is known; specifically, it is known if the passenger air bag had been sled-certified or if it had been certified to meet the 30 mph barrier test with unrestrained dummies. The age and restraint system of the child passenger is also known. Every SCI case vehicle, by definition, was equipped with a frontal air bag at the right front seat.

SCI cases are sometimes not investigated until well after the time of the crash. Notification may not be immediate, or an investigation may be triggered by a review of the Fatality Analysis Reporting System (FARS) or the Crashworthiness Data System (CDS). These follow-up reviews, especially in FARS, help SCI obtain a census of low-to-moderate speed fatalities in the United States, at least the ones that happened on public roads.

Even with those delays, 98 percent of the child fatalities were investigated no more than two calendar years after the year of the crash. Let us limit the analysis to the 162 fatalities in calendar years 1990 (when SCI started) through 2003; reporting would have been essentially complete by the end of 2005.[57]

R.L. Polk's National Vehicle Population Profile (NVPP) provides corresponding counts of registration years for vehicles equipped with passenger air bags. That permits calculation of fatality rates per billion years. However, to make SCI compatible with Polk, it is necessary to exclude another 7 SCI cases, leaving 155:

- 1 SCI fatality in Puerto Rico (not part of NVPP)

- 2 SCI cases where MY > CY – e.g., an early 1997 car that crashed in 1996 (not included or not well estimated in NVPP)[58]

- 4 SCI cases where the make-model was not 100% equipped with passenger air bags in that model year (NVPP does not identify how many of the vehicles of a given make-model and model year were equipped with air bags; the analysis must be limited to model years where all vehicles of that make-model were equipped).

- 0 SCI cases of MY 1998 vehicles of make-models whose air bags were sled-certified in mid-year (whereas SCI investigations identify if the air bag was redesigned or not, NVPP does not identify what proportion were redesigned; the analysis must be limited to model years where all vehicles of a make-model were redesigned, or none).

The objectives of the analysis are to distinguish the effect of the 1998-1999 redesign from the effect of earlier technological changes in air bags and from behavioral changes such as moving children to the back seat. Vehicles equipped with passenger air bags are initially split into three groups, based on these air bags (1) model year 1989-1994, (2) model year 1995+ but not sled-certified, and (3) sled-certified. A change in fatality rates from (1) to (2) could indicate the effect of technological changes before the 1998-1999 redesign. Calendar years are likewise split into three groups: 1990-1994, 1995-1997 and 1998-2003. A change in fatality rates across calendar years could indicate a change in behavior, especially a movement of children to the back seat in the calendar years after the public information campaigns started.

Table 2-1a shows SCI fatality rates, per billion years, by vehicle group and calendar year. For example, vehicles of model years 1989-1994 equipped with passenger air bags experienced 5 SCI fatalities in calendar years 1990-1994, during which they accumulated 5,593,290 registration years, a rate of 894.9 fatalities per billion years. The rates in calendar years 1995-1997 were similar, both for these vehicles (700.3) and for later, MY 1995+ vehicles that were not yet sled-certified (956.3). But in calendar years 1998-2003, even these vehicles with original air bags had

---

[57] The primary analyses of this chapter will compare fatality rates in the same calendar years of vehicles with original and redesigned air bags. Incompleteness of reporting will not create a bias favoring the vehicles with redesigned air bags, because the completeness of the reporting depends on the calendar year of the crash (same for both groups), not the model year of the vehicle.

[58] NVPP says how many vehicles of model year X were registered on July 1 of calendar year Y. Early vehicles of the "next" model year (X = Y + 1) are entirely omitted in some years of NVPP, and even when included the count of vehicles registered on July 1 is a poor estimate of the registration years they accumulated. Strictly speaking, even for vehicles of the "current" model year (X = Y), the registration count for July 1 is not that accurate an estimate of the registration years accumulated by that MY in that CY, but we will accept it here because deleting all SCI cases with MY = CY would be too great a loss of data.

substantially lower fatality rates (200.1 and 383.1, respectively), and sled-certified vehicles, much lower than that: 61.6.

TABLE 2-1a

CHILD PASSENGERS AGE 0-12 YEARS
SCI FATALITIES PER BILLION VEHICLE YEARS

Calendar Years	Vehicles	SCI Fatalities	Vehicle Years	SCI Fatalities per Billion Years
1990-1994	MY 1989-1994	5	5,593,290	893.9
1995-1997	MY 1989-1994	13	18,563,078	700.3
	MY 1995+, not sled-cert.	47	49,149,506	956.3
1998-2003	MY 1989-1994	7	34,979,017	200.1
	MY 1995+, not sled-cert.	68	177,502,313	383.1
	Sled-certified	15	243,523,549	61.6

Table 2-1a shows little difference between CY 1990-1994 and 1995-1997, and between the two groups of vehicles that are not sled-certified (taking into account that rates can vary considerably given the small N of SCI fatalities). Therefore, it makes sense to combine the two CY and two vehicle groups, as in Table 2-1b:

TABLE 2-1b

CHILD PASSENGERS AGE 0-12 YEARS
SCI FATALITIES PER BILLION VEHICLE YEARS

Calendar Years	Vehicles	SCI Fatalities	Vehicle Years	SCI Fatalities per Billion Years
1990-1997	Not sled-certified	65	73,305,874	886.7
1998-2003	Not sled-certified	75	213,481,330	353.0
	Sled-certified	15	243,523,549	61.6

Even for the vehicles that are not sled-certified, the fatality rate dropped by 60 percent from 886.7 in CY 1990-1997 to 353.0 in CY 1998-2003. Because these are the same vehicles, the great reduction is due to behavioral changes, above all moving children to the back seat, and if children stayed in the front seat, at least moving the seat all the way back and increasing use of restraints. Even more dramatic is the 83 percent reduction for sled-certification in CY 1998-2003: from 353.0 without sled certification to 61.6 for sled-certified vehicles. This would appear to be the effect of redesigned air bags, not behavioral changes, because we are comparing the two groups of vehicles in the same calendar years. The combined effect of technological and behavioral changes is a 93 percent reduction in the SCI fatality rate, from 886.7 with original air bags before calendar year 1998 to 61.6 with sled-certified air bags in calendar years 1998-2003.

Table 2-2 compares SCI fatality rates separately for four age groups of children:

TABLE 2-2: CHILD PASSENGERS AGE 0-12 YEARS, BY AGE GROUP SCI FATALITIES PER BILLION VEHICLE YEARS

Calendar Years	Vehicles	SCI Fatalities	Vehicle Years	SCI Fatalities per Billion Years
INFANTS IN REAR-FACING SAFETY SEATS				
1990-1997	Not sled-certified	13	73,305,874	177.3
1998-2003	Not sled-certified	11	213,481,330	51.8
	Sled-certified	none	243,523,549	0.0
CHILDREN AGE 0-5 YEARS, NOT IN REAR-FACING SAFETY SEATS				
1990-1997	Not sled-certified	38	73,305,874	518.4
1998-2003	Not sled-certified	44	213,481,330	207.1
	Sled-certified	11	243,523,549	45.2
CHILDREN AGE 6-10 YEARS				
1990-1997	Not sled-certified	14	73,305,874	191.0
1998-2003	Not sled-certified	19	213,481,330	89.4
	Sled-certified	4	243,523,549	16.4
CHILDREN AGE 11-12 YEARS				
1990-1997	Not sled-certified	1	73,305,874	13.6
1998-2003	Not sled-certified	1	213,481,330	4.7
	Sled-certified	none	243,523,549	0.0

There are not enough SCI cases for statistically meaningful results for the 11-12 year-old group. In the three younger age groups, there are huge reductions in SCI fatalities with sled-certified air bags. It is especially encouraging that sled-certified air bags had not killed a single infant in a rear-facing safety seat as of June 1, 2006, while there were 11 fatalities during calendar years 1998-2003 in older vehicles, with similar exposure in registration years. That is certainly good news; NHTSA had cautiously not estimated any specific benefit here for redesigned air bags (see Section 1.4). For 0-5 year-old children not in rear-facing seats, the fatality rate in calendar years 1998-2003 was 78 percent lower with sled-certified air bags (45.2) than with original air bags (207.1); for 6-10 year-old children, the reduction with sled-certified air bags was 82 percent. Both results are even more favorable than the 43.5 percent reduction NHTSA had hoped for. Perhaps other design changes in addition to depowering, such as a shift to hybrid air bags are contributing to the effect.

Table 2-2 also indicates the important effect of behavioral changes. Even in vehicles that were not sled-certified, the fatality rate for infants dropped by 71 percent from CY 1990-1997 (177.3) to CY 1998-2003 (51.8). For 0-5 year-old children not in rear-facing seats, the corresponding reduction was 60 percent. For 6-10 year-old children, it was yet smaller, 53 percent. These fatality reductions closely correspond to the trends of more children sitting in the back seat during that time period: strongest for infants and gradually less strong for older children (see Section 1.5).

Table 2-3 separately analyzes the effects in passenger cars, pickup trucks, SUVs and vans. Because cars were equipped with passenger air bags well before most LTVs, they account for the vast majority of the vehicle years with air bags not yet sled-certified. Almost all pickup trucks in Table 2-3 (except a few full crew cabs) were equipped with on-off switches.

In all four types of vehicles, the SCI fatality rate in calendar years 1998-2003 was lower with sled-certified air bags than with original air bags. In passenger cars, the reduction was 78 percent (from 340.1 to 73.5). Pickup trucks have low absolute rates of SCI fatalities for two reasons: children are less likely to ride in pickup trucks than other vehicles, and correct use of on-off switches eliminated the risk of deploying air bags for most children.

In SUVs, based on limited data, the fatality reduction with sled-certified air bags is 93 percent. That is reassuring given the previous findings by the Children's Hospital of Philadelphia that redesigned air bags were not reducing injury risk in SUVs (see Section 1.6).

The results for vans are especially encouraging. Before sled certification, vans had the highest fatality rates (at least in part because they have so many child passengers). As of June 1, 2006, not a single child fatality in a van had been caused by sled-certified air bags.

TABLE 2-3: CHILD PASSENGERS AGE 0-12 YEARS, BY VEHICLE TYPE
SCI FATALITIES PER BILLION VEHICLE YEARS

Calendar Years	Vehicles	SCI Fatalities	Vehicle Years	SCI Fatalities per Billion Years
		PASSENGER CARS		
1990-1997	Not sled-certified	51	62,313,906	818.4
1998-2003	Not sled-certified	58	170,518,528	340.1
	Sled-certified	10	136,075,127	73.5
		PICKUP TRUCKS (with on-off switches)		
1990-1997	Not sled-certified	none	853,302	0.0
1998-2003	Not sled-certified	1	6,355,041	157.4
	Sled-certified	4	32,030,947	124.9
		SUVs		
1990-1997	Not sled-certified	1	3,644,836	274.4
1998-2003	Not sled-certified	5	17,450,734	286.5
	Sled-certified	1	49,943,252	20.0
		VANS		
1990-1997	Not sled-certified	13	6,493,830	2001.9
1998-2003	Not sled-certified	11	18,157,027	605.8
	Sled-certified	none	25,474,223	0.0

2.2   SCI fatality rates for drivers

Driver air bags were installed in large numbers of cars three years before passenger air bags. That provided a three-year head start for learning about injuries due to contact with air bags and introducing remedies. For drivers, the proportion of these injuries that were lethal was far smaller than for child passengers. Nevertheless (because every vehicle has a driver but not so many have child passengers in the front seat), by the end of calendar year 2005, the SCI file included records of 85 drivers with fatal injuries from air bags in otherwise survivable, low-to-moderate speed crashes (Delta V < 25 mph). All of the crashes occurred in calendar years 1990-2003. Fatality rates per billion years can be calculated with corresponding counts, from Polk's NVPP files, of registration years for vehicles equipped with driver air bags. As in Section 2.1, to make SCI compatible with Polk, it is necessary to exclude 4 SCI cases, leaving 81: 2 SCI cases where the make-model was not 100% equipped with driver air bags in that model year and 2 SCI cases of MY 1998 vehicles of make-models whose air bags were sled-certified in mid-year.

Here, too, the analysis objectives are to distinguish, to the extent allowed by the somewhat limited data, the effect of the 1998-1999 redesign from the effect of earlier technological changes in air bags and from behavioral changes such as sitting farther away from the steering wheel. Similar to Section 2.1, vehicles equipped with driver air bags are initially split into three groups: (1) model year 1989-1994, (2) model year 1995+ but not sled-certified, and (3) sled-certified. Calendar years are likewise split into three groups: 1990-1994, 1995-1997 and 1998-2003.

Table 2-4 shows SCI driver fatality rates per billion years, by vehicle group and calendar year. Unlike the situation with children (Table 2-1a), driver fatality rates were already decreasing before the 1998-1999 redesign. For example, in calendar years 1995-1997, MY 1989-1994 vehicles experienced 288.0 SCI fatalities per billion years, but MY 1995+ vehicles that were not yet sled-certified, only 149.7, a 48 percent reduction. Similarly, in CY 1998-2003, the MY 1989-1994 vehicles had a fatality rate of 104.9, but MY 1995+ vehicles that were not yet sled-certified, only 57.0, a 46 percent reduction. In other words, there is a consistent reduction from MY 1989-1994 to the MY 1995+ air bags that were not yet sled certified. That would appear to be the effect of technological improvements even before the 1998-1999 redesign, such as tethering the air bags and reducing the rearward extent of their deployments (see Section 1.4).

TABLE 2-4

DRIVERS: SCI FATALITIES PER BILLION VEHICLE YEARS

Calendar Years	Vehicles	SCI Fatalities	Vehicle Years	SCI Fatalities per Billion Years
1990-1994	MY 1989-1994	19	51,551,083	368.6
1995-1997	MY 1989-1994	21	72,917,529	288.0
	MY 1995+, not sled-cert.	10	66,817,798	149.7
1998-2003	MY 1989-1994	14	133,501,278	104.9
	MY 1995+, not sled-cert.	13	228,209,084	57.0
	Sled-certified	4	234,454,019	17.1

Upon sled certification, the fatality rate plunged again, to 17.1. That is a 70 percent drop from the 57.0, in the same calendar years, in the MY 1995+ vehicles immediately before sled certification. It is a cumulative 84 percent reduction from 104.9, the fatality rate with original MY 1989-1994 air bags in calendar years 1998-2003. The results are consistent with NHTSA's prediction (see Section 1.4) that depowering "could reduce a large portion, but not all of these fatalities" on the driver's side.

Table 2-4 also shows a trend of lower fatality rates in later calendar years, even for the same vehicles. For example, the fatality rate in MY 1989-1994 vehicles fell from 368.6 in CY 1990-1994 to 288.0 in CY 1995-1997 to 104.9 in CY 1998-2003. Hopefully, the trend reflects growing public awareness of the risk of short drivers from air bags and response to publicity campaigns urging drivers to sit at least 10 inches from the air bag. However, some of it could also be due to less complete notification of SCI events in later years as the underlying problem diminished; unlike the child passenger fatalities, SCI does not review all FARS cases of driver fatalities and has less assurance that the cases on file are, in fact, a census of low-to-moderate speed fatalities.[59]

Combining the model-year and calendar-year effects, the drop from 368.6 with MY 1989-1994 vehicles in CY 1990-1994 to 17.1 with sled-certified vehicles in CY 1998-2003 is a 95 percent reduction in the SCI fatality rate.

Table 2-5 shows quite similar fatality reductions in passenger cars and LTVs upon sled-certification.

TABLE 2-5: DRIVERS, BY VEHICLE TYPE
SCI FATALITIES PER BILLION VEHICLE YEARS

Calendar Years	Vehicles	SCI Fatalities	Vehicle Years	SCI Fatalities per Billion Years
PASSENGER CARS				
1998-2003	Not sled-certified	20	260,959,260	76.6
	Sled-certified	2	128,105,739	15.6
PICKUP TRUCKS, SUVs and VANS				
1998-2003	Not sled-certified	7	100,751,102	69.5
	Sled-certified	2	106,348,280	18.8

2.3   SCI fatality rates for adult and teenage passengers

The risk from air bags to adult and teenage passengers is substantially lower than for child passengers. The numbers are also lower than for drivers, to a large extent because there has been less exposure. By the end of calendar year 2005, the SCI file included records of 13 right-front passengers age 13 years or more with fatal injuries from air bags in low-to-moderate speed crashes (Delta V < 25 mph). Fatality rates per billion registration years during CY 1990-2003 can be calculated with SCI and Polk data. To make SCI compatible with Polk, we exclude one

---

[59] However, underreporting would probably not bias the results of analyses that compare two different types of vehicles in the same calendar years – e.g., the fatality reduction for sled-certified vehicles vs. earlier vehicles in CY 1998-2003 – because the completeness of the reporting more likely depends on the calendar year of the crash (same for both groups) than on the model year of the vehicle.

SCI case of a crash in CY 2004 and one SCI case of a MY 1998 car of a make-model whose air bags were sled-certified in mid-year. Only a rudimentary analysis is possible with 11 SCI cases: Table 2-6, like Table 2-1b combines all vehicles that were not sled-certified and all calendar years before 1998.

TABLE 2-6

ADULT AND TEENAGE PASSENGERS (AGE 13 YEARS AND UP)
SCI FATALITIES PER BILLION VEHICLE YEARS

Calendar Years	Vehicles	SCI Fatalities	Vehicle Years	SCI Fatalities per Billion Years
1990-1997	Not sled-certified	6	73,305,874	81.8
1998-2003	Not sled-certified	3	213,481,330	14.1
	Sled-certified	2	243,523,549	8.2

Table 2-6, although based on small numbers of SCI cases, shows a fatality reduction for sled-certified vehicles, plus a reduction in fatality rates over time, not unlike the corresponding trends for child passengers and drivers. Specifically, the observed fatality rate for sled-certified vehicles in CY 1998-2003 is 42 percent lower than for vehicles that were not sled-certified. It is 90 percent lower than the fatality rate in CY 1990-1997.

CHAPTER 3

EFFECT OF REDESIGNED AIR BAGS
ON DRIVERS' OVERALL FATALITY RISK IN FRONTAL CRASHES

3.0   Summary

Drivers' fatality risk in frontal crashes remained unchanged after air bags were sled-certified (95 percent confidence bounds range from a 5 percent reduction to a 4 percent increase in fatality risk). In other words, the great life-saving benefits of first-generation driver air bags were preserved in their entirety. There was a small but not statistically significant increase in fatality risk for unrestrained drivers, offset by a non-significant reduction for belted drivers, resulting in zero net effect for all drivers. However, certain subsets of unrestrained drivers did experience significant increases in fatality risk with sled-certified or depowered air bags: drivers of passenger cars, especially if they were < 30 years old or > 6' tall; female, ≤ 5'3" and ≤ 125 pound drivers. Analyses are based on 1993-2004 Fatality Analysis Reporting System (FARS) data.

Even though the preceding chapter showed a 70 percent reduction, upon sled-certification, of fatalities caused by air bags in low-to-moderate speed crashes ("SCI fatalities"), that benefit makes an imperceptible contribution to the overall effect because SCI fatalities accounted for well under 1 percent of all frontal driver fatalities in vehicles equipped with original air bags.

3.1   Methods to analyze the effect of redesigned air bags

NHTSA's four statistical analyses of crash experience with air bags, published in 1992, 1996, 1999 and 2001, each used the same two methods to compute fatality-reducing effectiveness in frontal crashes – i.e., the fatality reduction for the original air bags of the early to mid-1990s relative to not having an air bag at all.[60]

One method is based on a distinctive characteristic of air bags: they are primarily designed for action in frontal crashes. In a crash where no impact has a frontal force component, air bags are unlikely to deploy. They can be assumed to have essentially no effect, positive or negative. These fatalities in non-frontal crash involvements are a control group.

In the Fatality Analysis Reporting System (FARS), we may define non-frontal crash involvements to include first-event rollovers and non-collisions, plus purely side or rear impacts (initial <u>and</u> principal impact locations between 3:00 and 9:00[61]). "Frontal" crashes have initial <u>or</u>

---

[60] Kahane, C.J., *Lives Saved by the Federal Motor Vehicle Safety Standards and Other Vehicle Safety Technologies, 1960-2002,* NHTSA Technical Report No. DOT HS 809 833, Washington, 2004, pp. 108-110; *Evaluation of the Effectiveness of Occupant Protection, Interim Report,* NHTSA Report No. DOT HS 807 843, 1992, pp. 21-25; Kahane, C.J., *Fatality Reduction by Air Bags,* NHTSA Technical Report No. DOT HS 808 470, Washington, 1996, pp. 7-12; *Fourth Report to Congress – Effectiveness of Occupant Protection Systems and Their Use,* NHTSA Report No. DOT HS 808 919, Washington, 1999, pp. 8-11; *Fifth/Sixth Report to Congress – Effectiveness of Occupant Protection Systems and Their Use,* NHTSA Report No. DOT HS 809 442, Washington, 2001, pp. 6-11.

[61] Or one of them between 3:00 and 9:00 and the other at an unknown/non-reported location.

principal impact between 11:00 and 1:00 (excluding first-event rollovers and non-collisions). Slightly frontal impacts with initial or principal impact at 10:00 or 2:00, but no impact at 11-1:00 are withheld from the analysis.[62] The dataset from NHTSA's 2001 study was based on CY 1986-2000 FARS data and limited to selected make-models that received driver (or dual) air bags at some point, but were always equipped with 3-point belts. Driver fatalities in cars of the last 3 model years before air bags are compared to fatalities in cars of the first 3 model years with driver (or dual) air bags:

	Frontal (11-1:00) Fatalities	Non-Frontal Fatalities	Frontal/Non-Frontal Risk Ratio
Driver seat without air bag	4,775	2,932	1.628
Driver seat equipped with air bag	2,865	2,269	1.263

Drivers of cars equipped with air bags experienced a 22 percent reduction of frontal fatalities:

$$1 - [(2,865/2,269) / (4,775/2,932)] = .2247$$

The fatality reduction is statistically significant, as evidenced by a Chi-square (χ^2) of 48.40. χ^2 must exceed 3.84 for statistical significance at the .05 level. The 95 percent confidence bounds for the effectiveness extend from 17 to 28 percent.[63]

In the preceding example, the method was used to estimate the effectiveness of original air bags relative to no air bags at all. But it applies equally well to estimating the reduction (or increase) in frontal fatalities for redesigned air bags relative to original air bags. It will be the principal technique in this chapter. In the above table, the first row would be "driver seat with original (i.e., not yet redesigned) air bag" and the second row would be "driver seat with redesigned air bag." No news would be good news. If redesigned air bags had the same risk ratio as original air bags, it would mean that redesigned air bags reduce frontal fatality risk by 22 percent relative to no air bags at all, the same as the original air bags.

For unbiased results with this method, data must be carefully selected. The vehicles with redesigned and original air bags should be, to the extent feasible, identical in all other features

---

62

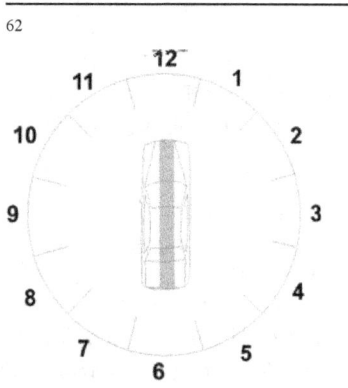

[63] Computed by the RELRISK option of the FREQ procedure in SAS® software.

(except for the change in the air bags). That includes features potentially affecting fatality risk in non-frontal crashes, because those crashes are used as the control group – e.g., side air bags, electronic stability control. The analysis should include only a limited number of model years before and after sled certification, because the ratio of frontal to non-frontal crashes can change as a vehicle ages, or from year to year.[64]

Double-pair comparison, the other method used for estimating fatality reduction of original air bags vs. no air bags, generally cannot be used to evaluate redesigned air bags.[65] It worked for original air bags because a large number of make-models were initially equipped with only a driver air bag for several years before 1994. The driver's seats changed from no air bags to air bag-equipped. The right-front passenger seat stayed the same: no air bag. That allowed double-pair comparison, with the right front passenger as the control group. The dataset from NHTSA's 2001 study has the following distribution of fatalities in 11-1:00 frontal impacts where the driver's and right-front seats were both occupied; and the driver, or the RF passenger, or possibly both died:

Driver's Seat	RF Passenger Seat	Driver Fatalities	RF Passenger Fatalities	Driver/RF Risk Ratio
No air bag	No air bag	11,894	12,493	0.952
Air-bag equipped	No air bag	1,658	2,224	0.746

Just as in the preceding method, drivers of cars equipped with air bags experienced a 22 percent reduction of frontal fatalities:

$$1 - [(1,658/2,224) / (11,894/12,493)] = .2170$$

This method cannot be used to evaluate redesigned air bags because, in the vast majority of make-models, driver and passenger air bags were sled-certified and/or depowered at the same time. Even in the models where it was not simultaneous, one followed the other in a year or less. There is no substantial group of vehicles where the driver air bag changed while the passenger bag stayed the same for several years before and after the change in the driver air bag. It is regrettable that this method cannot be used here, because it would have allowed including a wider range of data (because the ratio of driver to right-front passenger fatalities is rather invariant over time, or with vehicle age, or with the introduction of other safety devices that have similar effects for drivers and passengers). We will be able to use it in Chapter 5, analysis of child passengers, after we have demonstrated that the effect of redesigned air bags for drivers, if any, is negligible relative to the effect for child passengers.

A secondary method that will be used throughout this report is to compute fatality rates in frontal crashes per million vehicle registration years, using fatality counts from FARS and registration data from R.L. Polk's National Vehicle Population Profile (NVPP). Fatality rates are compared

---

[64] Kahane (1996), pp. 7-9 and 40-41.
[65] Evans, L., "Double Pair Comparison - A New Method to Determine How Occupant Characteristics Affect Fatality Risk in Traffic Crashes," *Accident Analysis and Prevention*, Vol. 18, June 1986, pp. 217-227.

before and after redesign. The method is similar to the analysis of SCI fatality rates in Chapter 2 as well as Braver's study of 2000-2002 FARS data.[66] Here, too, however, we must limit ourselves to a narrow range of model years and to make-models that did not change any feature that could affect fatality risk in frontal crashes (except for the change in the air bags). Otherwise, the effects of those features and/or the change in fatality rates with vehicle age could overshadow the generally small effect of redesigned air bags. (However, because we are only computing frontal fatality rates, we need not worry about features designed for non-frontal crashes, such as side impact protection.) Fortunately, enough years of FARS data have accumulated that we can be quite selective and still have a large N of cases.

3.2   Basic analyses of sled-certified air bags and depowered air bags

Definitions of "redesigned" air bags: Throughout this report, the primary definition of a "redesigned" air bag is a **sled-certified** air bag. The objective is to compare drivers' fatality risk immediately before vs. after sled certification. NHTSA has fairly complete tables of when each make-model's air bags were initially sled-certified on the driver's side.[67]

However, as discussed in Section 1.4, manufacturers had the option to sled-certify even if the air bags remained unchanged. Conversely, they might have redesigned air bags to some extent even before the sled-test option became available. A more selective definition of a "redesigned" air bag is a **depowered** air bag, based on the detailed performance specifications up to model year 1998, furnished to NHTSA by the manufacturers as confidential business information in response to the agency's Information Request (IR). We will say an air bag was depowered if the IR indicates:

- A reduction in the scaled peak pressure (see Section 1.5), or

- A reduction in the scaled rise rate, or

- Other evidence of redesign to make the air bag less aggressive, such as a shift from a pyrotechnic to a hybrid inflator or a reduction in the volume or the rearward extent of the air bag.

Sled certification and depowering usually coincided. The vast majority of make-models were sled-certified in the beginning or middle of model year 1998, and for most of these models NHTSA received IR data indicating they depowered at that same time. The exceptions are;

- A few make-models that sled-certified in 1998 but, according to the IR depowered in 1996 or 1997 and remained unchanged in 1998. These models would be included in both analyses, but the "sled-certified" analysis would compare before-1998 and 1998+, while the "depowered" analysis would compare before vs. after the depowering.

---

[66] Braver, E.R., Kyrychenko, S.Y., and Ferguson, S.A., "Driver Mortality in Frontal Crashes: Comparison of Newer and Older Air Bag Designs," *Traffic Injury Prevention*, Vol. 6, March 2005, pp. 24-30.

[67] *National Automotive Sampling System – Crashworthiness Data System – 2000 Coding and Editing Manual*, NHTSA, Washington, 2000, pp. 721-733, access from http://www-nrd nhtsa.dot.gov/departments/nrd-30/ncsa/AvailInf html .

- A few make-models that sled-certified in 1998, but the IR suggests the air bags remained unchanged in 1998 and in preceding years. These models are included in the "sled-certified" analysis and omitted in the "depowered" analysis.

- A larger number of models where there are no IR data at all, or at least not for the sled-certified air bags. Included are all make-models that sled-certified in 1999 or later (because the IR data end in 1998). These models, likewise, are included in the "sled-certified" analysis and omitted in the "depowered" analysis.

Because the "depowered" analysis is more selective than the "sled-certified" analysis, we might expect its results to be consistent with and, on the whole, slightly stronger than the corresponding results for the "sled-certified" analysis.

FARS data preparation – sled-certified air bags: The basic FARS analysis compares the ratio of frontal to non-frontal driver fatalities immediately before vs. after sled certification. The analysis is based on FARS data from calendar years 1994-2004. In the data preparation, all definitions of make-models are based on decoding the VIN. Since 1991, NHTSA staff has maintained a series of VIN analysis programs for use in evaluations. The programs are available to the public.

The goal in selecting vehicles for the analysis is to make the group of vehicles with sled-certified air bags as similar as possible to the group with original air bags (except for the change in air bags). To begin with, the analysis is limited to make-models produced both before and after sled certification. At a maximum, we include the last three model years before sled-certification and the first three sled-certified years. For example, if the entire 1998 model year is sled-certified, we include up to 1995-1997 and 1998-2000.

- If sled-certification begins in mid-year and the VIN identifies what vehicles are sled-certified, include five model years – e.g., for a make-model sled-certified in mid-1998, include 1996-2000, and identify from the VIN which of the 1998s are sled-certified.

- If sled-certification begins in mid-year and the VIN does not identify what vehicles are sled-certified, exclude that year and include three model years before and after it – e.g., for a make-model sled-certified in mid-1998, include 1995-1997 and 1999-2001, and exclude all of the 1998s.

- Exclude make-models where NHTSA has no information on when they sled-certified.

Of course, if a make-model began production less than three years before sled-certification, or ended less than three years afterwards, we are immediately limited to the years it was produced. "Starting or ending production" would also include replacing a make-model with a radically different vehicle of the same name (e.g., Volkswagen Beetle and New Beetle), but not the customary restyling where a vehicle remains in the same functional class with fairly small changes in wheelbase or appearance.

The maximum span of six model years is likewise abbreviated if there are any of the following major changes in the vehicle's safety systems during that time:

- Exclude any vehicle not equipped with driver air bags. If the make-model was initially equipped with driver air bags less than three years before sled certification, exclude the model years without air bags.

- Exclude any vehicle equipped with dual-stage driver air bags. If the make-model upgraded to dual-stage air bags less than three years after sled certification, exclude the model years with dual-stage air bags.

- Omit the few make-models that up-powered at the same time they sled-certified. A model "up-powered" if the IR shows an increase in the scaled peak pressure and/or the scaled rise rate (see Section 1.5), without a decrease in either parameter or any other evidence of redesign (such as a shift from pyrotechnic to hybrid inflators).

- Safety-belt pretensioners might generally affect fatality risk (of belted drivers) and specifically influence a belted driver's interaction with the air bag. Limit the span of model years to a group that is homogeneous with respect to pretensioners.[68] For example, if a make-model was sled-certified in 1998, and pretensioners were introduced in:
    - 1996 or 1997 – exclude the years before pretensioners.
    - 1998 – exclude the make-model entirely.
    - 1999 or 2000 – exclude the years with pretensioners.

- The range of model years must likewise be abbreviated upon introduction of major safety systems that affect fatality risk in the control group of non-frontal crashes, specifically:
    - Initial certification to FMVSS No. 214 (Side Impact Protection) – if in 1996 or 1997, exclude any model years before FMVSS No. 214 certification.
    - Side air bags – the vehicles (of a particular make-model) included in the analysis must be homogeneous – i.e., none of them have side air bags, or all of them have torso air bags, or all of them have torso plus head air bags. If side air bags were introduced (or upgraded from torso to torso-plus-head) at the same time as sled certification, exclude the make-model entirely.
    - Electronic Stability Control (ESC) – exclude all vehicles equipped with ESC or that may have been equipped with optional ESC. If ESC was introduced at the same time as sled certification, exclude the make-model entirely.

Finally, model years are deleted from any make-model if that is necessary to **balance** the sample between original and sled-certified air bags – i.e., to assure more or less the same make-model mix in the sample with original air bags as in the sample with sled-certified air bags. Thus, for example, if for any of the preceding reasons, we can only use the first model year of sled-certified vehicles for a particular make-model, we will also use only the last model year before sled certification. (However, because the earlier vehicles are on the road longer and have

---

[68] However, we do not abbreviate the model years upon introduction of load limiters for safety belts because (1) the effect is probably smaller; (2) this would exclude a large proportion of the data; and (3) it is not certain when load limiters were introduced in many of the make-models.

accumulated more data, we can allow one extra model year with sled-certified air bags if we are limited to just one or two years with original air bags).[69]

Basic FARS analysis of sled-certified air bags: Table 3-1, based on FARS data for calendar years 1994-2004, shows that drivers in frontal crashes had virtually identical fatality risk with original and sled-certified air bags.

TABLE 3-1

DRIVERS: OVERALL EFFECT OF SLED-CERTIFIED AIR BAGS
(1994-2004 FARS, as evidenced by ratio of frontal to non-frontal fatalities)

	Frontal (11-1:00) Fatalities	Non-Frontal Fatalities	Frontal/Non-Frontal Risk Ratio
Not sled-certified	9,273	7,169	1.293
Sled-certified	7,629	5,911	1.291

Drivers of vehicles equipped with sled-certified air bags experienced a 0.2 percent reduction of frontal fatalities, compared to original air bags in the same make-models:

$$1 - [(7,629/5,911) / (9,273/7,169)] = .0022$$

The fatality reduction is not statistically significant, as evidenced by a χ^2 of 0.01. The 95 percent confidence bounds for the effect extend from –4 to +5 percent. Because original air bags reduce fatality risk in frontal crashes by 22 percent relative to no air bags at all, sled-certified air bags also reduce fatality risk in frontal crashes by 22 percent relative to no air bags at all.

Basic FARS analysis of depowered air bags: The data preparation for the analysis of depowered air bags is nearly identical. As discussed above, the dividing line between "before" and "after" is when make-models depowered (as early as 1996), not necessarily when they sled-certified (1998 at the earliest). If there is no IR, or the IR does not show a depowering, the model is excluded. The few models that up-powered are again omitted. Table 3-2 is based on FARS data for calendar years 1993-2004. It shows that drivers in frontal crashes had virtually identical fatality risk with original and depowered air bags.

---

[69] See also Kahane (1996), pp. 7-9.

TABLE 3-2

DRIVERS: OVERALL EFFECT OF DEPOWERED AIR BAGS
(1993-2004 FARS, as evidenced by ratio of frontal to non-frontal fatalities)

	Frontal (11-1:00) Fatalities	Non-Frontal Fatalities	Frontal/Non-Frontal Risk Ratio
Not depowered	7,479	5,476	1.366
Depowered	5,995	4,440	1.350

Drivers of vehicles equipped with depowered air bags experienced a 1 percent reduction of frontal fatalities, compared to original air bags in the same make-models:

$$1 - [(5,995/4,440) / (7,479/5,476)] = .0114$$

The fatality reduction is not statistically significant, as evidenced by a χ^2 of 0.19. The 95 percent confidence bounds for the effect extend from –4 to +6 percent. Because original air bags reduce fatality risk in frontal crashes by 22 percent relative to no air bags at all, depowered air bags reduce fatality risk in frontal crashes by 23 percent relative to no air bags at all.

FARS/Polk analysis of sled-certified air bags: The secondary analysis method is to compare drivers' frontal fatality rates per million vehicle registration years immediately before vs. after sled certification. The analysis is based on FARS data and R.L. Polk's National Vehicle Population Profile (NVPP) from calendar years 1996-2004. Whereas NVPP data do not include the actual VIN, NHTSA staff has developed a series of programs that use the VIN-derived variables on NVPP to define make-models corresponding exactly to the FARS data, allowing NVPP to be merged with FARS. NVPP specifies the number of vehicles registered as of July 1 of every calendar year. "Frontal" crashes, as before, have initial or principal impact between 11:00 and 1:00 (excluding first-event rollovers and non-collisions).

Here, too, the goal is to make the group of vehicles with sled-certified air bags as similar as possible to the group with original air bags (except for the change in air bags). As in the preceding analysis, we include at most the last three model years before sled-certification and the first three sled-certified years. We also use all of the additional filters from the preceding analysis – e.g., exclude vehicles without air bags, exclude dual-stage air bags, do not allow a change in the status of pretensioners – except the requirements that the included vehicles be homogeneous with respect to FMVSS No. 214 certification and side air bags. Because this analysis is limited to frontal impacts only, it does not matter if the vehicles have acquired technology designed primarily for side impact protection.[70]

We will also impose one additional filter: the analysis is limited to vehicles at least a year old at the time of the crash (CY > MY). NVPP says how many vehicles N of model year X were registered on July 1 of calendar year Y. When Y > X, N is also an excellent estimate of how

---

[70] However, because ESC can prevent frontal as well as non-frontal impacts, we will continue to exclude make-models in model years with standard or optional ESC.

many registration years vehicles of model year X accumulated in calendar year Y (because, except for a small proportion of vehicles that were scrapped or registered late, the vehicles registered on July 1 were there the whole year). But for new vehicles (Y = X) or early introductions of the next model year (Y = X – 1), the count of vehicles registered on July 1 is not that accurate an estimate of the registration years they accumulated (because new vehicles may be initially registered at any time during the year, including after July 1) – at least, not accurate enough for this analysis, which is looking for small changes in fatality rates. Also, if a make-model was sled-certified in mid-model year that model year needs to be excluded from the analysis because the Polk data do not distinguish between the original and sled-certified air bags of the same model year.

Table 3-3 shows little change in drivers' fatality rates, in frontal crashes, per million registration years, following sled-certification. The fatality rate with sled-certified air bags (54.43) is 2 percent lower than with original air bags (55.60).

TABLE 3-3: DRIVERS, FRONTAL FATALITIES PER MILLION VEHICLE YEARS
BEFORE AND AFTER SLED CERTIFICATION
(1996-2004 FARS)

	Frontal (11-1:00) Fatalities	Vehicle Years	Frontal Fatalities per Million Years
Not sled-certified	9,441	169,796,288	55.60
Sled-certified	6,687	122,863,415	54.43

The 2 percent reduction is not statistically significant, based on the following test: there were 133 individual make-models that had non-zero fatality rates for original and/or sled-certified air bags. In 68 of these make-models, the fatality rate was lower with sled-certified than with original air bags; in 65 it was higher. That is not a significant departure from a 50-50 split.[71]

FARS/Polk analysis of depowered air bags: Similarly, when the data are limited to make-models known to have been depowered and the analysis focuses on when these models depowered (rather than on when they sled-certified), Table 3-4 shows little change in drivers' fatality rates in frontal crashes after depowering. The fatality rate with depowered air bags (56.14) is 2 percent lower than with original air bags (57.29).

---

[71] $p = 68/133 = .5113$; $s = \sqrt{(.25/133)} = .0434$; $z = (p - .5)/s = 0.26$.

TABLE 3-4: DRIVERS, FRONTAL FATALITIES PER MILLION VEHICLE YEARS BEFORE AND AFTER DEPOWERING (1995-2004 FARS)

	Frontal (11-1:00) Fatalities	Vehicle Years	Frontal Fatalities per Million Years
Not depowered	7,841	136,868,024	57.29
Depowered	5,287	94,169,640	56.14

The 2 percent reduction is not statistically significant, based on the following test: there were 84 individual make-models that had non-zero fatality rates for original and/or depowered air bags. In 44 of these make-models, the fatality rate was lower with depowered than with original air bags; in 40 it was higher. That is not a significant departure from a 50-50 split.[72]

Tables 3-1 – 3-4 all suggest that the redesign of air bags in 1998-1999 had little or no net effect on the overall fatality risk of drivers in frontal crashes. At least, the effect was too small to be discerned in these analyses of crash data – and by now, the vehicles have been on the road for quite a few years and have accumulated a large portion of their eventual exposure. No news is good news: the redesigned air bags are about equally effective for drivers as the first-generation air bags. The remainder of this chapter will look at the effect of redesigned air bags for certain subgroups of occupants, vehicles and crashes. Some of the detailed analyses will show a negative effect for redesigned air bags relative to original air bags, and some of those effects may even be "real," as evidenced by their statistical significance, consistency with results of similar analyses, and consistency with intuition. But the reader should keep in mind that all the results add up to zero or something close to it. If redesigned air bags provide less protection than original air bags for certain drivers, they must have made up for it by enhancing protection for other drivers, because, overall, they are equally effective as original air bags.

The results might, at first glance, seem inconsistent with the findings of Chapter 2, especially Table 2-4. They indicated a 70 percent reduction, upon sled-certification, of fatalities caused by air bags in low-to-moderate speed crashes ("SCI fatalities"). But Table 1-1 showed that SCI fatalities accounted for only 0.27 percent of all frontal fatalities with original air bags. A 70 percent reduction of SCI fatalities corresponds to a $70 \times .0027 = 0.19$ percent overall fatality reduction. The impact of a 0.19 percent effect is completely lost in the statistical "noise" of analyses of such as Table 3-1 that estimate the overall effectiveness of sled-certified air bags in frontal crashes to a precision of ± 4.5 percent. Even in the most vulnerable subgroup of drivers – unrestrained females $\leq$ 5'3" and age 70 or older – only 4 percent of frontal fatalities were SCI cases (Table 1-8) and the SCI reduction would be lost in the "noise" level of an analysis of the overall effect in that subgroup.

---

[72] $p = 44/84 = .5238$; $s = \sqrt{(.25/84)} = .0546$; $z = (p - .5)/s = 0.44$.

3.3   Unrestrained vs. belted drivers

Unrestrained and belted drivers should be analyzed separately. Engineering intuition suggests that unrestrained occupants could be less protected if air bags are depowered: in severe frontals unrestrained occupants need all the energy absorption they can get from the air bag because they have failed to use safety belts to absorb part of the load (see Section 1.2). NHTSA's testing of 1996 generally supported that, showing consistently increased risk with depowered air bags for unrestrained (correctly positioned) drivers and passengers, but mixed results for belted occupants (see Section 1.4).

Crash data are also consistent with intuition, up to a point. The upper part of Table 3-5 shows that unrestrained drivers had 5 percent higher fatality risk in frontal crashes with sled-certified air bags than with original air bags. The increase is not statistically significant, as evidenced by a χ^2 of 2.80. By contrast, the lower half of Table 3-5 shows that belted drivers experienced a 5 percent reduction of frontal fatality risk with sled-certified air bags. The reduction is also non-significant, as evidenced by a χ^2 of 1.81. (Because drivers with unknown belt use are not included in Table 3-5, the numbers do not add up to the totals in Table 3-1.)

TABLE 3-5

UNRESTRAINED VS. BELTED DRIVERS: EFFECT OF SLED-CERTIFIED AIR BAGS
(1994-2004 FARS, as evidenced by ratio of frontal to non-frontal fatalities)

	Frontal (11-1:00) Fatalities	Non-Frontal Fatalities	Risk Ratio	Reduction (%)
	UNRESTRAINED DRIVERS			
Not sled-certified	5,121	3,997	1.281	
Sled-certified	4,067	3,009	1.352	− 5
	BELTED DRIVERS			
Not sled-certified	3,348	2,726	1.228	
Sled-certified	2,957	2,532	1.168	5

Nevertheless, a CATMOD analysis of the data in Table 3-5 does show a statistically significant interaction between belt use, sled certification and the probability that a fatality is frontal, as evidenced by a χ^2 of 4.45 for the coefficient of that interaction term.[73] In other words, the effect of sled-certification on frontal fatality risk is significantly different for unrestrained and belted drivers: worse for the unrestrained driver than for the belted driver.

---

[73] *SAS/STAT® User's Guide, Version 6, Fourth Edition*, Volume 2, SAS Institute, Cary, NC, 1989, pp. 405-517. The dependent variable is the impact type (frontal or non-frontal). The independent variables are belt use and sled certification.

A similar analysis compares the effect of depowered air bags for unrestrained and belted drivers. The effects are directionally the same, but weaker. Unrestrained drivers had 3 percent higher fatality risk in frontal crashes with depowered air bags than with original air bags. Belted drivers experienced a 4 percent reduction with depowered air bags. Neither change is statistically significant. Furthermore, in the CATMOD analysis, the χ^2 for the interaction term is only 1.50, also non-significant. (As stated above, if the effects were genuine, they ought to be stronger, not weaker, for the subgroup of make-models known to have depowered.)

3.4   Direct (12:00) vs. oblique (11 or 1:00) frontal impacts

Intuition also suggests that the disadvantage of depowered air bags, if there is one, might be greatest in straight-ahead, directly forward impacts that call upon the highest energy absorption for the air bag. The capacity of the air bag may be less relevant in an oblique, glancing impact, where occupants are more likely to engage with various other components of the vehicle.

Here, too, the results from crash data are directionally consistent with intuition. Let us define a "directly frontal 12:00 impact" as one where both the initial and principal impact point is 12:00 (they are likely one and the same impact; also include cases where one of these impacts is 12:00 and the other is unknown). "All other 11-1:00 impacts" include all the other frontals in Table 3-1; they include cases where the initial and/or principal impact is 11:00 or 1:00, plus cases where one of the impacts is 12:00 and the other is not frontal at all. Table 3-6 computes fatality risk for each of these subgroups of frontals relative to the same, usual control group, non-frontal crash involvements. With sled-certified air bags, there is a 3 percent increase in directly frontal fatalities. The increase is not statistically significant, as evidenced by a χ^2 of 1.03. As we know, the overall effect of sled-certified air bags is zero; therefore, there must be a positive effect in the oblique frontal crashes to offset the negative at 12:00, and there is: a 7 percent reduction that is statistically significant, as evidenced by a χ^2 of 4.37.[74]

The corresponding analysis of depowered vs. original air bags produces similar results: a non-significant 2 percent fatality increase in the direct 12:00 impacts ($\chi^2 = 0.68$), offset by a significant 9 percent fatality reduction in the other frontal impacts ($\chi^2 = 5.16$). There is no obvious, intuitive reason that sled-certified or depowered air bags should be more effective than original air bags in oblique frontals. Perhaps it is due to modifications over the years in air bags that are not directly related to depowering, or even due to modifications in other parts of the vehicle. In any case, the net effect of redesigned vs. original air bags in all frontal crashes adds up to zero.

---

[74] Moreover, the simple 2x2 table created by the four numbers in the left column of Table 3-6 (direct vs. other frontals) has a significant χ^2 of 8.83, indicating that sled-certified air bags are significantly more effective in the other frontals than in the direct 12:00 impacts.

TABLE 3-6

DIRECT VS. OBLIQUE FRONTAL IMPACTS: EFFECT OF SLED-CERTIFIED AIR BAGS
(1994-2004 FARS, as evidenced by ratio of frontal to non-frontal fatalities)

	Direct (12:00) Frontal Fatalities	Non-Frontal Fatalities	Risk Ratio	Reduction (%)
Not sled-certified	6,061	7,169	.845	
Sled-certified	5,152	5,911	.872	− 3

	Other (11-1:00) Frontal Fatalities	Non-Frontal Fatalities	Risk Ratio	Reduction (%)
Not sled-certified	3,212	7,169	.448	
Sled-certified	2,477	5,911	.419	7

3.5   Analyses for subgroups of drivers, vehicles and crashes

Table 3-7 computes the fatality reductions for redesigned air bags, relative to original air bags in the same make-models, for various subgroups of drivers, vehicles and crash types. For each analysis, Table 3-7 shows the percent fatality reduction, computed relative to the control group of non-frontal impacts, as in Tables 3-1, 3-2, 3-5 and 3-6, and its associated χ^2.

For example, the first row and first two columns of Table 3-7 repeat the basic analysis of sled-certified air bags in Table 3-1: the overall frontal fatality reduction for sled-certified air bags was zero percent (because sled-certified and original air bags had the same ratio of frontal to non-frontal fatalities), and the χ^2 for the 2x2 Table 3-1 was 0.01. Similarly, the next two columns of the first row of Table 3-7 repeat the basic analysis of depowered air bags from Table 3-2: a 1 percent fatality reduction for depowered air bags, with a χ^2 of 0.19. The right half of Table 3-7 computes effectiveness for just the directly frontals with initial and principal 12:00 impacts. As in Table 3-6, the overall effect for sled-certified air bags is –3 percent, with a χ^2 of 1.39, and for depowered air bags, –2 percent, with a χ^2 of 0.68.

Positive results (fatality reduction for redesigned relative to original air bags) are printed in black; so are findings of no change, because no news is also good news. Negative results are red. Statistically significant positive results are highlighted in blue, significant negatives in yellow.

The left column of numbers is the most important one in Table 3-7. These are the fatality reductions of all 11-1:00 frontal crashes with sled-certified air bags, where we have the most data. Because the overall effect is zero, there is understandably a mix of positive and negative findings, but **none of them is statistically significant**. The other three findings in each row (namely, when the analysis is limited to make-models known to have depowered, or to directly frontal 12:00 impacts, or both) are basically a consistency check on the 11-1:00, sled-certified finding.

TABLE 3-7: ALL DRIVERS, PERCENT FATALITY REDUCTION FOR REDESIGNED RELATIVE TO ORIGINAL AIR BAGS

	In 11, 12 and 1:00 Frontals				In 12:00 Frontals			
	All Sled Cert Models		Depowered Models Only		All Sled Cert Models		Depowered Models Only	
	% Red.	χ^2	% Red.	χ^2	% Red.	χ^2	% Red.	χ^2
OVERALL	none	.01	1	.19	-3	1.39	-2	.68
BELT USE								
Not belted	-5	2.80	-3	.70	-10	7.93	-7	3.08
Belted	5	1.81	4	.81	3	.47	1	.03
AGE								
13-29	-2	.32	none	.01	-5	1.39	-2	.21
30-55	3	.89	4	.99	1	.03	1	.07
56-69	2	.06	2	.08	-3	.19	-5	.32
70+	-3	.24	-2	.06	-7	.90	-7	.65
GENDER								
Male	2	.43	1	.07	-1	.06	-2	.42
Female	-5	1.12	none	.00	-10	4.07	-5	.76
DRIVER HEIGHT								
Up to 5'3"	-7	1.12	-3	.18	-13	2.77	-8	.72
5'4" to 6'	3	.80	3	.66	2	.37	1	.13
Over 6'	-3	.15	2	.10	-8	1.01	-3	.13
DRIVER WEIGHT								
Up to 125 pounds	-9	1.24	1	.01	-10	1.23	-1	.01
126-199 pounds	4	1.26	5	1.55	1	.09	3	.47
200 pounds and up	2	.13	3	.26	none	.00	none	.00
VEHICLE TYPE								
Passenger car	-4	1.31	-7	2.96	-9	6.10	-14	8.56
LTV	3	.97	7	4.12	2	.18	5	1.92
Pickup truck	2	.33	2	.33	none	.00	none	.00
SUV	1	.02	11	2.67	-1	.04	11	1.82
Van	-12	1.05	-13	.74	-10	.66	-16	.98
CRASH TYPE								
Single-vehicle	-1	.06	1	.10	-2	.46	1	.03
Multivehicle	2	.40	2	.18	-2	.39	-5	1.24

The next two rows of Table 3-7 compare results for unrestrained and belted drivers. The left columns repeat Table 3-5, showing, for unbelted drivers, a 5 percent increase of frontal fatalities with sled-certified air bags, and for belted drivers, a 5 percent reduction. In fact, all four estimates for unrestrained drivers are negative, but only one of them, the effect of sled-certified air bags on 12:00 impacts, is statistically significant. All four estimates for belted drivers are positive, but none of them are statistically significant.

The age of the driver has little interaction with the effect of sled-certified air bags. All estimates are close to zero and non-significant. Original air bags were effective for drivers of all ages[75], and redesigned air bags are more or less equally effective.

The effect of redesigned air bags is consistently the least favorable for three groups that largely coincide: females, people ≤ 5'3" tall, and people ≤ 125 pounds. Conversely, drivers of medium height (5'4" – 6') and weight (126-199 pounds) have the most positive results. But, at least in Table 3-7, the differences are small and not statistically meaningful. We'll look at these factors again for certain subgroups, especially unbelted drivers, in the next tables.

The effect of redesigned air bags tends to be negative in passenger cars and vans, positive in pickup trucks and SUVs. Differences are small in the "main" analyses of sled-certified vehicles and 11-1:00 frontals, though there are three significant results in the other analyses. But this analysis method definitely does not produce the large, statistically significant [and, frankly, inexplicable] differences among vehicle types reported in an earlier publication (discussed in Section 1.6).[76]

The effect of redesigned air bags is about the same in single- and multivehicle crashes.

Table 3-8 computes the fatality reductions for redesigned air bags, relative to original air bags, for unrestrained drivers. It analyzes subgroups of occupants, vehicles and crash types. Because the overall effect for unrestrained drivers is negative, it is not surprising to see negatives in most of the subgroups. Nevertheless, relatively few negatives are statistically significant, and most of them are in the 12:00 impacts.

For all unrestrained drivers, the fatality increase in sled-certified models is a non-significant 5 percent in 11-1:00 impacts and a significant 10 percent in just the 12:00 impacts. But when the data are limited to models known to have been depowered, the effect is weaker, not stronger, and it is not significant. As in Table 3-7, there is little interaction with drivers' age.

However, Table 3-8 shows consistently negative results for unrestrained females, drivers ≤ 5'3" and drivers ≤ 125 pounds, achieving statistical significance on three analyses. That is surprising at first glance, for we had assumed that unrestrained large people have the greatest need for high-capacity air bags. Perhaps smaller occupants are loading primarily the lower portion of the air bags, and when the air bags are depowered these occupants "punch through" and engage the steering assembly with their lower thorax and abdomen. Perhaps depowered air bags take a bit longer to inflate, allowing short occupants who sit up close to reach the steering wheel before full inflation of the air bag.

---

[75] Kahane (2004), p. 111.
[76] Braver (2005).

TABLE 3-8: UNRESTRAINED DRIVERS
PERCENT FATALITY REDUCTION FOR REDESIGNED RELATIVE TO ORIGINAL AIR BAGS

| | In 11, 12 and 1:00 Frontals | | | | In 12:00 Frontals | | | |
| | All Sled Cert Models | | Depowered Models Only | | All Sled Cert Models | | Depowered Models Only | |
	% Red.	χ^2	% Red.	χ^2	% Red.	χ^2	% Red.	χ^2
ALL UNRESTRAINED	-5	2.80	-3	.70	**-10**	**7.93**	-7	3.08
AGE								
13-29	-7	1.73	-5	.71	-11	3.68	-8	1.42
30-55	-3	.31	none	.00	-8	2.14	-4	.35
56-69	-12	.84	-6	.25	-13	1.15	-13	.92
70+	-5	.19	-3	.04	-11	.67	-10	.42
GENDER								
Male	-3	.59	-2	.34	-5	1.97	-5	1.14
Female	-13	3.73	-6	.53	**-26**	**10.41**	-17	3.43
DRIVER HEIGHT								
Up to 5'3"	-16	2.16	-9	.48	**-33**	**6.21**	-20	1.70
5'4" to 6'	-3	.45	-2	.17	-4	.95	-3	.37
Over 6'	-6	.39	2	.06	-15	1.79	-8	.42
DRIVER WEIGHT								
Up to 125 pounds	**-25**	**3.88**	-23	2.31	-28	3.74	-28	2.66
126-199 pounds	-1	.03	4	.46	-4	.57	3	.23
200 pounds and up	7	**.87**	8	.95	3	.12	3	.07
VEHICLE TYPE								
Passenger car	-10	3.45	-17	6.35	**-22**	**12.35**	**-30**	**15.29**
LTV	-2	.27	4	.71	-3	.42	4	.53
Pickup truck	none	.00	none	.00	-1	.01	-1	.01
SUV	-12	2.61	6	.40	-15	3.00	7	.45
Van	-24	1.53	-30	1.61	-24	1.26	-35	1.74
CRASH TYPE								
Single-vehicle	-4	1.18	-2	.22	-7	2.69	-3	.44
Multivehicle	-7	1.51	-5	.59	**-15**	**5.04**	-14	3.48

Table 3-8 also shows consistently negative results for passenger cars and vans, achieving statistical significance in three of the four analyses of passenger cars. To the extent that cars are lighter and less rigid than pickup trucks and SUVs, they experience a higher proportion of exceptionally severe frontal impacts, where unrestrained drivers may especially need the extra energy absorption of original air bags.

Concentrating on the bad news, Table 3-8a focuses on unrestrained drivers of passenger cars. Here, there are quite a few significant negatives, but most of them are in the 12:00 impacts. In the "main" analyses of 11-1:00 impacts for sled-certified models, there is still not a single significant negative.

Nevertheless, focusing on this group of often severe frontals, especially on the 12:00 impacts that more directly involve interaction with the air bag, reveals trends consistent with intuitive expectations. For the first time, there appears to be some interaction with driver age. Young drivers, who have the most need and the least vulnerability to aggressive air bags, show the most negative effect for redesigned air bags. Smaller drivers (females, $\leq 5'3''$, ≤ 125 pounds) continue to be worse off with redesigned than with original air bags – but now, for the first time, drivers over 6' tall are also clearly worse off, at least in 12:00 crashes. Tall drivers may be concentrating load on the upper portion of the air bags, and "punching through" depowered air bags to engage the steering assembly.

The consistent negatives in single-vehicle crashes hint that unrestrained drivers who are jostled to one side or another during the off-road excursion before an impact with a fixed object are obtaining less benefit from smaller, less aggressive, less protrusive redesigned air bags. (Another factor may be that young – i.e., tall – male drivers are overrepresented in single-vehicle crashes.)

Even though Table 3-8a contains rather negative results that appear to confirm some of the concerns about depowered air bags, there are three good reasons not to be alarmed:

- The overall effect of redesigned air bags is zero; whatever was negative in Table 3-8a is offset somewhere else by positives.

- First-generation air bags are exceptionally effective for unrestrained drivers, reducing fatality risk in 12:00 impacts by 34 percent.[77] Even if redesigned air bags are not as effective for unrestrained drivers as original air bags, they are still effective relative to no air bags at all.

- There is an exceedingly simple and increasingly popular way for drivers to avoid any of these problems:

BUCKLE UP

---

[77] Kahane (2004), p. 111.

TABLE 3-8a: UNRESTRAINED DRIVERS OF PASSENGER CARS
PERCENT FATALITY REDUCTION FOR REDESIGNED RELATIVE TO ORIGINAL AIR BAGS

	In 11, 12 and 1:00 Frontals				In 12:00 Frontals			
	All Sled Cert Models		Depowered Models Only		All Sled Cert Models		Depowered Models Only	
	% Red.	χ^2	% Red.	χ^2	% Red.	χ^2	% Red.	χ^2
ALL UNRESTRAINED CAR DRIVERS	-10	3.45	-17	6.35	-22	12.35	-30	15.29
AGE								
13-29	-13	2.67	-25	5.79	-25	7.70	-43	12.21
30-55	-7	.64	-12	1.28	-17	2.76	-17	2.07
56-69	-8	.16	-7	.10	-15	.54	-21	.70
70+	1	.00	1	.00	-16	.64	-21	.85
GENDER								
Male	-8	1.56	-14	3.28	-19	6.65	-29	9.77
Female	-14	2.05	-22	3.21	-27	5.89	-33	5.44
DRIVER HEIGHT								
Up to 5'3"	-9	.33	-14	.56	-21	1.51	-24	1.35
5'4" to 6'	-4	.33	-8	1.16	-9	1.83	-16	3.26
Over 6'	-23	1.60	-15	.54	-63	7.32	-62	5.31
DRIVER WEIGHT								
Up to 125 pounds	-27	2.37	-38	2.68	-23	1.31	-34	1.75
126-199 pounds	-1	.02	-2	.06	-10	1.38	-11	.99
200 pounds and up	2	.01	3	.03	-13	.66	-13	.51
CRASH TYPE								
Single-vehicle	-13	3.75	-27	9.06	-23	8.07	-37	12.57
Multivehicle	-3	.14	-2	.03	-16	2.90	-17	2.25

Table 3-9 computes the fatality reductions for redesigned air bags, relative to original air bags, for belted drivers. Because the overall effect for belted drivers is slightly positive, so are the effects in most of the subgroups. Only three are statistically significant: the effects in all frontals for LTVs in general (sled-certified and depowered models) and SUVs in particular (sled-certified models only). The combination of safety belt use and relatively low-severity crashes in these heavy, rigid vehicles reduces the need for the energy-absorbing capacity of original air bags while perhaps enhancing the benefit of contact with a softer air bag.

On the other hand, Table 3-9 shows consistently negative (although non-significant) effects for drivers over 6' tall or weighing ≥ 200 pounds. Perhaps these large occupants "punch through" the top of the air bag even when they wear safety belts. Another possible explanation is that they sit so far back from the steering wheel that, when belted, they make little contact with the smaller, less protrusive redesigned air bags and, as a result, might receive less benefit than with first-generation air bags.

TABLE 3-9: BELTED DRIVERS
PERCENT FATALITY REDUCTION FOR REDESIGNED RELATIVE TO ORIGINAL AIR BAGS

	In 11, 12 and 1:00 Frontals				In 12:00 Frontals			
	All Sled Cert Models		Depowered Models Only		All Sled Cert Models		Depowered Models Only	
	% Red.	χ^2	% Red.	χ^2	% Red.	χ^2	% Red.	χ^2
ALL BELTED	5	1.81	4	.81	3	.47	1	.03
AGE								
13-29	5	.53	8	1.11	2	.07	7	.67
30-55	10	3.19	8	1.35	10	2.64	5	.49
56-69	6	.34	3	.07	-1	.01	-4	.13
70+	-7	.64	-8	.64	-9	.83	-10	.74
GENDER								
Male	5	1.13	2	.08	2	.20	-2	.12
Female	3	.34	5	.64	2	.12	4	.24
DRIVER HEIGHT								
Up to 5'3"	-5	.27	-1	.01	-3	.08	-3	.05
5'4" to 6'	8	3.45	6	1.22	8	2.44	4	.56
Over 6'	-12	.99	-10	.50	-15	1.24	-10	.41
DRIVER WEIGHT								
Up to 125 pounds	7	.36	20	2.53	5	.15	19	1.64
126-199 pounds	8	2.62	6	.95	7	1.60	4	.40
200 pounds and up	-13	1.50	-10	.81	-15	1.79	-10	.70
VEHICLE TYPE								
Passenger car	-3	.33	-6	1.02	-4	.63	-8	1.42
LTV	**12**	**4.83**	**13**	**4.61**	9	2.25	9	1.82
Pickup truck	4	.28	4	.28	-1	.00	-1	.00
SUV	**20**	**5.32**	20	3.20	19	3.56	17	1.70
Van	2	.01	13	.53	1	.01	10	.27
CRASH TYPE								
Single-vehicle	2	.17	3	.24	4	.29	6	.67
Multivehicle	6	1.95	4	.57	3	.26	-2	.12

54

CHAPTER 4

EFFECT OF REDESIGNED AIR BAGS ON ADULT AND TEENAGE PASSENGERS' OVERALL FATALITY RISK IN FRONTAL CRASHES

4.0   Summary

The fatality risk of adult and teenage passengers in the right-front seat did not change significantly in frontal crashes after air bags were sled-certified (95 percent confidence bounds range from a 13 percent reduction to a 5 percent increase in fatality risk with sled-certified air bags, relative to original air bags). In other words, the great life-saving benefits of first-generation passenger air bags for occupants age 13 years and up were preserved in their entirety. Analyses are based on 1994-2004 Fatality Analysis Reporting System (FARS) data.

Even though Chapter 2 showed a 42 percent reduction, upon sled-certification, of fatalities caused by air bags in low-speed crashes ("SCI fatalities"), that benefit makes an imperceptible contribution to the overall effect because SCI fatalities accounted for well under 1 percent of adult and teenage passenger fatalities in frontal crashes in vehicles equipped with original air bags.

4.1   Basic analyses of sled-certified air bags and depowered air bags

The methods for estimating the effect of redesigned air bags for adult and teenage right-front passengers are almost identical to the analysis of drivers, as defined in Sections 3.1 and 3.2. Before we compare the fatality risk with redesigned and original air bags, let us review the fatality reduction in frontal crashes for those original air bags of the early to mid-1990s relative to not having an air bag at all. The major finding of earlier analyses is that original air bags have great life-saving benefits for passengers age 13 years and up, but increase risk for young children. Let us focus, in this chapter, on the teenage and adult passengers.[78]

One method for evaluating original air bags, which will also be the primary method for redesigned air bags is to compare the ratio of frontal to non-frontal fatalities "before" and "afterwards." The fatalities in non-frontal crashes are a control group, because air bags are unlikely to deploy in those crashes.

In FARS, as in Section 3.1, non-frontal crash involvements include first-event rollovers and non-collisions, plus purely side or rear impacts (initial <u>and</u> principal impact locations between 3:00 and 9:00[79]). "Frontal" crashes have initial <u>or</u> principal impact between 11:00 and 1:00 (excluding first-event rollovers and non-collisions). Slightly frontal impacts with initial or principal impact at 10:00 or 2:00, but no impact at 11-1:00 are withheld from the analysis. The

---

[78] Kahane, C.J., *Lives Saved by the Federal Motor Vehicle Safety Standards and Other Vehicle Safety Technologies, 1960-2002,* NHTSA Technical Report No. DOT HS 809 833, Washington, 2004, pp. 112-113; Kahane, C.J., *Fatality Reduction by Air Bags,* NHTSA Technical Report No. DOT HS 808 470, Washington, 1996, pp. 14-18.
[79] Or one of them between 3:00 and 9:00 and the other at an unknown/non-reported location.

analysis is based on CY 1986-2000 FARS data and limited to selected make-models that received passenger air bags at some point, but were always equipped with 3-point belts.[80] Right-front (RF) passenger fatalities, age 13 years and up, in cars of the last 3 model years before passenger air bags are compared to fatalities in cars of the first 3 model years with passenger air bags:

	Frontal (11-1:00) Fatalities	Non-Frontal Fatalities	Frontal/Non-Frontal Risk Ratio
RF seat without air bag	864	699	1.236
RF seat equipped with air bag	595	704	.845

Adult and teenage right-front passengers of cars equipped with air bags experienced a 32 percent reduction of frontal fatalities:

$$1 - [(595/704) / (864/699)] = .3162$$

The fatality reduction is statistically significant, as evidenced by a Chi-square (χ^2) of 25.48. χ^2 must exceed 3.84 for statistical significance at the .05 level. The 95 percent confidence bounds for the effectiveness extend from 21 to 41 percent.[81]

Double-pair comparison also provides an estimate of fatality reduction for original passenger air bags vs. no air bags, because a large number of make-models were initially equipped with only a driver air bag for several years before 1994.[82] The RF seat changed around 1994 from no air bags to air bag-equipped. The driver's seat stayed the same: air-bag equipped. That allows double-pair comparison, with the driver as the control group. The CY 1986-2000 FARS data have the following distribution of fatalities in 11-1:00 frontal impacts where the driver's and right-front seats were both occupied; and the driver, or the RF passenger, or possibly both died:

80

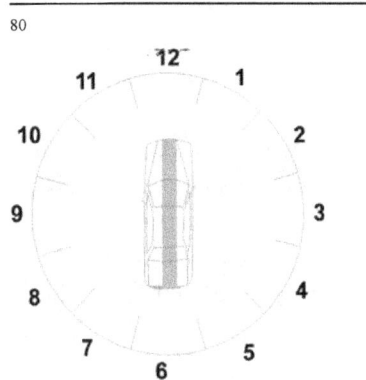

---

[81] Computed by the RELRISK option of the FREQ procedure in SAS® software.
[82] Evans, L., "Double Pair Comparison - A New Method to Determine How Occupant Characteristics Affect Fatality Risk in Traffic Crashes," *Accident Analysis and Prevention*, Vol. 18, June 1986, pp. 217-227.

RF Passenger Seat	Driver's Seat	RF Passenger Fatalities	Driver Fatalities	RF/Driver Risk Ratio
No air bag	Air-bag equipped	2,180	1,578	1.381
Air-bag equipped	Air-bag equipped	2,120	2,068	1.025

Adult and teenage right-front passengers of cars equipped with air bags experienced a 26 percent reduction of frontal fatalities:

$$1 - [(2,120/2,068) / (2,180/1,578)] = .2579$$

The two analyses produce somewhat different effectiveness estimates for passenger air bags, 32 and 26 percent. However, both estimates are higher than the corresponding results for driver air bags (22 percent in each analysis; see Section 3.1). The average is 29 percent. In the analysis of redesigned air bags, even no news is good news. If redesigned air bags had the same risk ratio as original air bags, it would mean that redesigned air bags reduce frontal fatality risk by 29 percent relative to no air bags at all, the same as the original air bags.

As in Chapter 3, the first analysis method – comparing the ratio of frontal to non-frontal fatalities – is also the principal method for analyzing redesigned air bags. (Double-pair comparison, on the other hand, cannot be used because driver and passenger air bags were sled-certified and/or depowered at the same time in the vast majority of make-models.) The secondary analysis method will be to compare, for redesigned and original air bags, the fatality rates in frontal crashes per million vehicle registration years, based on fatality counts from FARS and registration data from R.L. Polk's National Vehicle Population Profile (NVPP).

Definitions of "redesigned" air bags: As in Section 3.2, the primary definition of a "redesigned" air bag is a **sled-certified** air bag. NHTSA has fairly complete tables of when each make-model's air bags were initially sled-certified on the RF passenger's side.[83] Almost all make-models certified for the passenger at the same time as the driver.

A secondary, more selective definition of a "redesigned" air bag is a **depowered** air bag, based on what the manufacturers furnished to NHTSA as confidential business information in response to our Information Request (IR). We will say an air bag was depowered if the IR indicates a reduction in the scaled peak pressure, or a reduction in the scaled rise rate, or other evidence of redesign to make the air bag less aggressive, such as a shift from a pyrotechnic to a hybrid inflator. Whereas the majority of make-models sled-certified and depowered both the driver and passenger air bags in the beginning or middle of model year 1998, manufacturers occasionally depowered one side while keeping the other unchanged (or even up-powering), or depowered the two sides in different years. There are also quite a few models where there are no IR data at all, or where the IR data ended before any depowering. These models are included in the "sled-certified" analysis and omitted in the "depowered" analysis.

---

[83] *National Automotive Sampling System – Crashworthiness Data System – 2000 Coding and Editing Manual*, NHTSA, Washington, 2000, pp. 721-733, access from http://www-nrd.nhtsa.dot.gov/departments/nrd-30/ncsa/AvailInf.html .

FARS data preparation: The basic FARS analysis compares the ratio of frontal to non-frontal fatalities of right-front passengers age 13 years and older, immediately before vs. after sled certification. The analysis is based on FARS data from calendar years 1994-2004. The data preparation is generally the same as for drivers (see Section 3.2), including:

- Definitions of make-models based on decoding the VIN.

- Limiting the data to at most the last three model years before sled-certification and the first three sled-certified years.

- Further limiting the span of model years by:
  - Excluding any vehicle not equipped with passenger air bags.
  - Excluding any vehicle equipped with dual-stage passenger air bags.
  - Assuring all the included vehicles from a given make-model are homogeneous with regard to pretensioners, FMVSS No. 214 certification (Side Impact Protection), side air bags and Electronic Stability Control (ESC).

- Excluding make-models:
  - Where NHTSA has no information on when they sled-certified.
  - That up-powered at the same time they sled-certified, as evidenced by an increase in the scaled peak pressure and/or the scaled rise rate, without a decrease in either parameter or any other evidence of redesign.
  - That introduced passenger air bags, or side air bags, or pretensioners or ESC at the same time they sled-certified. Indeed, many pickup trucks and full-sized vans introduced passenger air bags in 1998, sled-certified from the start; because these models never had passenger air bags that were not sled-certified, they are excluded from the study.

- **Balancing** the sample between original and sled-certified air bags by excluding "before" model years if the above factors resulted in an exclusion of "after" model years, and vice-versa. For example, if for any of the preceding reasons, we can only use the first model year of sled-certified vehicles for a particular make-model, we will also use only the last model year before sled certification.

One additional stipulation unique to passenger air bags: all vehicles of a given make-model must be homogeneous with regard to factory-installed on-off switches for the passenger air bags. Either they all have them (pickup truck models) or they do not (all other models). Pickup trucks without on-off switches, such as certain full crew cabs are excluded. We do not want to compare air bags that are "on" all the time to air bags that may be "off" part of the time.

Basic FARS analysis of sled-certified air bags: Table 4-1, based on FARS data for calendar years 1994-2004, shows that right-front passengers age 13 years and older in frontal crashes had about the same fatality risk with original and sled-certified air bags.

TABLE 4-1

RIGHT-FRONT PASSENGERS AGE 13 YEARS AND OLDER
OVERALL EFFECT OF SLED-CERTIFIED AIR BAGS
(1994-2004 FARS, as evidenced by ratio of frontal to non-frontal fatalities)

	Frontal (11-1:00) Fatalities	Non-Frontal Fatalities	Frontal/Non-Frontal Risk Ratio
Not sled-certified	1,687	1,860	.907
Sled-certified	1,509	1,744	.865

Adult and teenage RF passengers of vehicles equipped with sled-certified air bags experienced a 5 percent reduction of frontal fatalities, compared to original air bags in the same make-models:

$$1 - [(1,509/1,744) / (1,687/1,860)] = .0460$$

The fatality reduction is not statistically significant, as evidenced by a χ^2 of 0.94. The 95 percent confidence bounds for the effect extend from −5 to +13 percent. Because original air bags reduce fatality risk in frontal crashes by 29 percent relative to no air bags at all, sled-certified air bags may be reducing fatality risk in frontal crashes by 32 percent relative to no air bags at all.[84]

Basic FARS analysis of depowered air bags: The data preparation for the analysis of depowered air bags is nearly identical, except the dividing line between "before" and "after" is when make-models depowered (as early as 1997), not necessarily when they sled-certified (1998 at the earliest). If there is no IR, or if the IR does not show a depowering, the model is excluded. Table 3-2 is based on FARS data for calendar years 1994-2004. Adult and teenage RF passengers in frontal crashes had virtually identical fatality risk with original and depowered air bags.

TABLE 4-2

RIGHT-FRONT PASSENGERS AGE 13 YEARS AND OLDER
OVERALL EFFECT OF DEPOWERED AIR BAGS
(1994-2004 FARS, as evidenced by ratio of frontal to non-frontal fatalities)

	Frontal (11-1:00) Fatalities	Non-Frontal Fatalities	Frontal/Non-Frontal Risk Ratio
Not depowered	1,225	1,359	.901
Depowered	1,083	1,217	.890

---

[84] $1 - [(1-.29)x(1-.046)]$.

Passengers of vehicles equipped with depowered air bags experienced a 1 percent reduction of frontal fatalities, compared to original air bags in the same make-models:

$$1 - [(1,083/1,217) / (1,225/1,359)] = .0128$$

The fatality reduction is not statistically significant, as evidenced by a χ^2 of 0.05. The 95 percent confidence bounds for the effect extend from –12 to +10 percent. Because original air bags reduce fatality risk in frontal crashes by 29 percent relative to no air bags at all, depowered air bags may reduce fatality risk in frontal crashes by 30 percent relative to no air bags at all.

FARS/Polk analysis of sled-certified air bags: The secondary analysis method is to compare RF passengers' frontal fatality rates per million vehicle registration years immediately before vs. after sled certification. The analysis is based on FARS data and R.L. Polk's National Vehicle Population Profile (NVPP) from calendar years 1996-2004. NVPP specifies the number of vehicles registered as of July 1 of every calendar year. "Frontal" crashes have initial or principal impact between 11:00 and 1:00 (excluding first-event rollovers and non-collisions).

As in the preceding analysis and in Section 3.2, the data include at most the last three model years before sled-certification and the first three sled-certified years. The data are subjected to all of the additional filters from the preceding analysis except the requirements that the included vehicles be homogeneous with respect to FMVSS No. 214 certification and side air bags.[85] The data are further limited to vehicles at least a year old at the time of the crash (CY > MY), because from that age onward NVPP vehicle counts are accurate estimates of accumulated registration years.

Table 4-3 shows adult and teenage RF passengers' fatality rates, in frontal crashes, per million registration years, before and after sled-certification. The fatality rate with sled-certified air bags (12.49) is 9 percent lower than with original air bags (13.80).

TABLE 4-3

RIGHT-FRONT PASSENGERS AGE 13 YEARS AND OLDER, FRONTAL FATALITIES
PER MILLION VEHICLE YEARS, BEFORE AND AFTER SLED CERTIFICATION
(1996-2004 FARS)

	Frontal (11-1:00) Fatalities	Vehicle Years	Frontal Fatalities per Million Years
Not sled-certified	1,787	129,517,139	13.80
Sled-certified	1,330	106,465,063	12.49

The 9 percent reduction is not statistically significant, based on the following test: there were 120 individual make-models that had non-zero fatality rates for original and/or sled-certified air bags.

---

[85] Because this analysis is limited to frontal impacts only, it does not matter if the vehicles have acquired technology designed primarily for side impact protection.

In 67 of these make-models, the fatality rate was lower with sled-certified than with original air bags; in 53 it was higher. That is not a significant departure from a 50-50 split.[86]

FARS/Polk analysis of depowered air bags: Similarly, when the data are limited to make-models known to have been depowered and the analysis focuses on when these models depowered (rather than on when they sled-certified), Table 4-4 shows little change in passengers' fatality rates in frontal crashes after depowering. The fatality rate with depowered air bags (12.89) is 6 percent lower than with original air bags (13.75).

TABLE 4-4

RIGHT-FRONT PASSENGERS AGE 13 YEARS AND OLDER, FRONTAL FATALITIES
PER MILLION VEHICLE YEARS, BEFORE AND AFTER DEPOWERING
(1996-2004 FARS)

	Frontal (11-1:00) Fatalities	Vehicle Years	Frontal Fatalities per Million Years
Not depowered	1,328	96,560,353	13.75
Depowered	931	72,201,645	12.89

The 6 percent reduction is not statistically significant, based on the following test: there were 67 individual make-models that had non-zero fatality rates for original and/or depowered air bags. In 34 of these make-models, the fatality rate was lower with depowered than with original air bags; in 33 it was higher. That is not a significant departure from a 50-50 split.[87]

None of Tables 4-1 – 4-4 suggest that the redesign of air bags in 1998-1999 had much net effect on the overall fatality risk of adult and teenage RF passengers in frontal crashes. If there is an effect, it is too small to be discerned in these analyses of crash data – and by now, the vehicles have been on the road for quite a few years and have accumulated a large portion of their eventual exposure. No news is good news: the redesigned air bags are about equally effective for adult and teenage passengers as the first-generation air bags. The remainder of this chapter will look at the effect of redesigned air bags for certain subgroups of passengers, vehicles and crashes. The reader should keep in mind that all those detailed results add up to zero or something close to it. The reader should also keep in mind that fatalities caused by air bags in low-speed crashes ("SCI fatalities") accounted for less than 1 percent of all frontal fatalities to adult and teenage passengers with original air bags (see Table 1-1). Even if redesigned air bags were highly effective in reducing SCI fatalities (and Table 2-6 suggests they were) that effect would be lost in the statistical "noise" of analyses of such as Table 4-1 that estimate the overall effectiveness of sled-certified air bags in frontal crashes to a precision of $\pm$ 8 percent.

---

[86] p = 67/120 = .5583; s = $\sqrt{(.25/120)}$ = .0456; z = (p - .5)/s = 1.28.
[87] p = 34/67 = .5075; s = $\sqrt{(.25/67)}$ = .0611; z = (p - .5)/s = 0.12.

4.2   Unrestrained vs. belted passengers

Unrestrained and belted passengers should be analyzed separately. Engineering intuition and NHTSA's testing of 1996 suggested that, if anybody, unrestrained occupants would be more likely to lose benefits if air bags were depowered, because they need more energy absorption from the air bag (see Sections 1.2 and 1.4). Crash data, however, do not show increased fatality risk with redesigned air bags for unrestrained adult and teenage passengers. In fact, the upper part of Table 4-5 indicates that unrestrained passengers had 7 percent lower fatality risk in frontal crashes with sled-certified air bags than with original air bags. The reduction is not statistically significant, as evidenced by a χ^2 of 0.99. The lower half of Table 3-5 shows that belted passengers experienced a 5 percent reduction of frontal fatality risk with sled-certified air bags. This reduction is also non-significant, as evidenced by a χ^2 of 0.51. (Passengers with unknown belt use are not included in Table 4-5.)

TABLE 4-5

UNRESTRAINED VS. BELTED RIGHT-FRONT PASSENGERS AGE 13 YEARS AND
OLDER EFFECT OF SLED-CERTIFIED AIR BAGS
(1994-2004 FARS, as evidenced by ratio of frontal to non-frontal fatalities)

	Frontal (11-1:00) Fatalities	Non-Frontal Fatalities	Risk Ratio	Reduction (%)
UNRESTRAINED PASSENGERS				
Not sled-certified	800	827	.967	
Sled-certified	662	736	.899	7
BELTED PASSENGERS				
Not sled-certified	730	895	.816	
Sled-certified	690	890	.775	5

A CATMOD analysis of the data in Table 4-5 does show a statistically significant interaction between belt use, sled certification and the probability that a fatality is frontal, as evidenced by a χ^2 of 0.05 for the coefficient of that interaction term.[88] In other words, the effect of sled-certification on frontal fatality risk is about the same for unrestrained and belted passengers.

A similar analysis examines the effect of depowered air bags. Unrestrained passengers had 2 percent lower fatality risk in frontal crashes with depowered air bags than with original air bags. So did belted passengers. Neither reduction is statistically significant. In the CATMOD analysis, the χ^2 for the interaction term is 0.00, also non-significant.

---

[88] *SAS/STAT® User's Guide, Version 6, Fourth Edition*, Volume 2, SAS Institute, Cary, NC, 1989, pp. 405-517. The dependent variable is the impact type (frontal or non-frontal). The independent variables are belt use and sled certification.

These results contrast with the findings for drivers. Redesigned air bags had a negative observed effect for unrestrained drivers, but not for belted drivers (see Section 3.3). One possibility is that the difference observed between unrestrained and belted drivers was not "real." (It was statistically significant only in the three-way CATMOD analysis; the individual effects for unrestrained drivers and belted drivers were each not significantly different from zero.) Conversely, there could be a real effect for passengers similar to the effect for drivers, but the sample sizes for passengers are insufficient to detect it. A third possibility is that the effects are actually different for drivers and passengers: when drivers "punch through" lower-pressure air bags they appreciably increase their risk of injury from contacting the steering assembly, but passengers who punch through to the instrument panel do not increase their risk as much.

4.3   Direct (12:00) vs. oblique (11 or 1:00) frontal impacts

Intuition also suggests that the disadvantage of depowered air bags, if there is one, might be greater in straight-ahead impacts that call upon the highest energy absorption for air bags than in oblique, glancing impacts, where occupants are more likely to engage with various other components of the vehicle. Here, the results from crash data are directionally consistent with intuition (lower effectiveness at 12:00), but just barely (not a statistically meaningful difference in effectiveness). As in Section 3.4, "directly frontal 12:00 impacts" are limited to those where both the initial and principal impact point is 12:00. Table 4-6 computes fatality risk separately for the directly frontal impacts and for all other frontal (11-1:00) impacts relative to the same, usual control group: non-frontal crash involvements. With sled-certified air bags, there is a 4 percent fatality reduction in directly frontal fatalities. The reduction is not statistically significant, as evidenced by a χ^2 of 0.49. In the oblique frontal crashes the fatality reduction is 6 percent, also not significant, as evidenced by a χ^2 of 0.85.[89]

TABLE 4-6

DIRECT VS. OBLIQUE FRONTAL IMPACTS: EFFECT OF SLED-CERTIFIED AIR BAGS
(1994-2004 FARS, as evidenced by ratio of frontal to non-frontal fatalities)

	Direct (12:00) Frontal Fatalities	Non-Frontal Fatalities	Risk Ratio	Reduction (%)
Not sled-certified	1,084	1,860	.583	
Sled-certified	978	1,744	.561	4

	Other (11-1:00) Frontal Fatalities	Non-Frontal Fatalities	Risk Ratio	Reduction (%)
Not sled-certified	603	1,860	.324	
Sled-certified	531	1,744	.304	6

---

[89] Moreover, the simple 2x2 table created by the four numbers in the left column of Table 4-6 (direct vs. other frontals) has a significant χ^2 of 0.11, indicating that sled-certified air bags are about equally effective in the oblique frontals and in the direct 12:00 impacts.

The corresponding analysis of depowered vs. original air bags produces similar results: a non-significant 1 percent fatality reduction in the direct 12:00 impacts ($\chi^2 = 0.01$), and a non-significant 2 percent fatality reduction in the other frontal impacts ($\chi^2 = 0.08$).

For drivers, the contrast between direct and oblique frontal crashes was substantially stronger (see Section 3.4). Here, too, there are three possible explanations for the differences between drivers and passengers. Either the data for drivers exaggerate the difference between direct and oblique frontal crashes, or Table 4-6 understates it, or the effects actually diverge for drivers and passengers as a consequence of the different structures in front of the air bags.

4.4   Analyses for subgroups of passengers, vehicles and crashes

Table 4-7 computes the fatality reductions for redesigned air bags, relative to original air bags in the same make-models, for various subgroups of passengers, vehicles and crash types. For each analysis, Table 4-7 shows the percent fatality reduction and its associated χ^2, computed relative to the control group of non-frontal impacts, as in Tables 4-1, 4-2, 4-5 and 4-6. From left to right, each row shows the percentage fatality reduction in: (1) all 11-1:00 frontal fatalities upon sled certification (the most important estimate), (2) all frontal fatalities upon depowering, (3) directly frontal 12:00 fatalities upon sled certification and (4) directly frontal fatalities upon depowering.

Positive results (fatality reduction for redesigned relative to original air bags) are printed in black; so are findings of no change, because no news is also good news. Negative results are red. Statistically significant positive results are highlighted in blue, significant negatives in yellow.

The first three rows of Table 4-7 summarize the results so far in this chapter: redesigned air bags were about as effective as original air bags for all adult and teenage RF passengers, unrestrained as well as belted, in all frontals as well as in directly frontal 12:00 impacts. All the observed fatality reductions for redesigned air bags in these first three rows were positive or zero, but none were statistically significant.

In the remainder of Table 4-7, three rows attract attention:

- Passengers age 70 or older have consistently higher fatality risk with redesigned than with original air bags, although the difference is significant only in direct frontals, with depowered air bags. Otherwise, there is little interaction with the passenger's age or gender. (Passengers' height and weight are not reported on FARS.)

- The effect is consistently negative (although not statistically significant) in vans, whereas it is close to zero and often positive in passenger cars, pickup trucks and SUVs.

- Sled-certified air bags are significantly more effective than original air bags in single-vehicle crashes – i.e., frontal impacts with fixed objects.

Let us withhold comment until we see to what extent these effects persist when data on unrestrained and belted passengers are analyzed separately.

TABLE 4-7: ALL RIGHT-FRONT PASSENGERS AGE 13 YEARS AND OLDER
PERCENT FATALITY REDUCTION FOR REDESIGNED RELATIVE TO ORIGINAL AIR BAGS

	In 11, 12 and 1:00 Frontals				In 12:00 Frontals			
	All Sled Cert Models		Depowered Models Only		All Sled Cert Models		Depowered Models Only	
	% Red.	χ^2	% Red.	χ^2	% Red.	χ^2	% Red.	χ^2
OVERALL	5	.94	1	.05	4	.49	1	.01
BELT USE								
Not belted	7	.99	2	.08	7	.78	2	.04
Belted	5	.51	2	.09	2	.06	none	.00
AGE								
13-29	6	.75	11	1.60	8	.99	13	1.90
30-55	11	1.50	none	.00	12	1.40	5	.14
56-69	13	.80	16	1.01	2	.02	3	.02
70+	-15	1.51	-26	3.75	-15	1.30	-30	3.85
GENDER								
Male	5	.44	10	1.49	5	.33	12	1.90
Female	4	.44	-7	.77	3	.16	-11	1.33
VEHICLE TYPE								
Passenger car	5	.71	none	.00	5	.63	2	.05
LTV	1	.02	none	.00	-1	.02	-5	.18
Pickup truck	15	.74	2	.00	9	.19	-14	.19
SUV	none	.00	3	.03	-1	.01	-2	.02
Van	-28	1.53	-26	1.34	-28	1.41	-28	1.32
CRASH TYPE								
Single-vehicle	**15**	**5.14**	12	2.26	**16**	**4.39**	15	2.69
Multivehicle	-5	.64	-8	1.15	-7	.91	-12	1.61

Table 4-8 computes fatality reductions for redesigned air bags, relative to original air bags, for unrestrained passengers. Table 4-9 does the same for belted passengers. The negative effect for older passengers persists in both tables. The negative effect for vans is entirely limited to the belted passengers. The positive effect in single-vehicle crashes persists in both tables, but it is significant only for the unrestrained passengers, who also show negative effects in multivehicle crashes. Aside from these, Tables 4-8 and 4-9 do not show any significant or numerically large effects.

At first glance, the negative effect for passengers age 70 or older runs counter to the intuition that the frailest occupants should benefit most from less aggressive air bags. But that intuition would apply primarily to out-of-position occupants. For the large majority of correctly positioned occupants who do not interact with air bags until they are fully, or almost fully deployed, original air bags could be advantageous for the older and frailer occupant. If the occupant is unrestrained, it may help to have extra energy-absorbing capacity and prevent "punching through" to the instrument panel; if he or she is belted, the larger air bag may help reduce some of the loading on the safety belt. On the other hand, no corresponding effect was seen for older drivers, unrestrained or belted, in Section 3.5.

The observed negative effects in vans are not statistically significant; the limited data do not support a firm conclusion that fatality risk increased. Nevertheless, the results are intriguing because vans are precisely where Table 2-3 showed by far the highest rate of SCI fatalities to child passengers with original air bags and the lowest rate – namely, no fatalities at all – with sled-certified air bags. It is at least conceivable that passenger air bags were more intensively redesigned in vans than in other types of vehicles, resulting in exceptional success at protecting child passengers. Perhaps that had a cost of reducing protection to adults. Specifically, belted adults may have relatively less interaction with smaller, less protrusive redesigned air bags – and with less interaction, lower potential for benefits.

The enhanced effectiveness of redesigned air bags in single-vehicle crashes has no obvious explanation, especially considering that no corresponding effect was seen for drivers in Section 3.5. However, one characteristic of fixed objects such as trees or poles is that they are narrow; damage to the vehicle is deep, but the deceleration is relatively gradual. Depowered air bags are well suited to these conditions. By contrast, head-on collisions with another vehicle, unless there is much offset, result in abrupt decelerations similar to a barrier impact. They call for maximum energy absorption by the air bag when the occupant is unrestrained. Original air bags may be advantageous. Another possibility is that sensors might have improved over the model year 1995-2000 timeframe to more quickly detect and respond to collisions with fixed objects and deploy the bags earlier.

TABLE 4-8: UNRESTRAINED RIGHT-FRONT PASSENGERS AGE 13 YEARS AND OLDER
PERCENT FATALITY REDUCTION FOR REDESIGNED RELATIVE TO ORIGINAL AIR BAGS

	In 11, 12 and 1:00 Frontals				In 12:00 Frontals			
	All Sled Cert Models		Depowered Models Only		All Sled Cert Models		Depowered Models Only	
	% Red.	χ^2	% Red.	χ^2	% Red.	χ^2	% Red.	χ^2
ALL UNRESTRAINED	7	.99	2	.08	7	.78	2	.04
AGE								
13-29	9	.86	8	.51	10	.91	11	.72
30-55	13	1.14	none	.00	18	1.71	6	.12
56-69	-15	.24	-5	.03	-10	.11	-3	.01
70+	-19	.49	-28	.76	-45	1.96	-67	2.96
GENDER								
Male	2	.06	2	.03	5	.24	8	.40
Female	13	1.62	3	.05	10	.70	-6	.16
VEHICLE TYPE								
Passenger car	3	.08	-2	.03	1	.02	-4	.10
LTV	6	.24	none	.00	8	.30	2	.01
Pickup truck	19	.80	-10	.09	18	.61	-11	.08
SUV	-1	.00	-6	.06	-2	.01	-13	.20
Van	9	.08	7	.05	19	.32	16	.22
CRASH TYPE								
Single-vehicle	21	6.20	18	3.06	21	4.70	20	2.94
Multivehicle	-25	3.33	-29	3.13	-24	2.71	-32	3.06

TABLE 4-9: BELTED RIGHT-FRONT PASSENGERS AGE 13 YEARS AND OLDER
PERCENT FATALITY REDUCTION FOR REDESIGNED RELATIVE TO ORIGINAL AIR BAGS

| | In 11, 12 and 1:00 Frontals | | | | In 12:00 Frontals | | | |
| | All Sled Cert Models | | Depowered Models Only | | All Sled Cert Models | | Depowered Models Only | |
	% Red.	χ^2	% Red.	χ^2	% Red.	χ^2	% Red.	χ^2
ALL BELTED	5	.51	2	.09	2	.06	none	.00
AGE								
13-29	12	1.18	24	3.69	11	.73	24	2.59
30-55	12	.75	2	.02	12	.59	9	.21
56-69	18	.98	20	.99	none	.00	-5	.04
70+	-17	1.39	-32	3.84	-12	.58	-28	2.40
GENDER								
Male	10	.77	19	2.39	6	.20	19	1.77
Female	2	.03	-8	.69	-1	.00	-12	.98
VEHICLE TYPE								
Passenger car	6	.54	2	.05	5	.26	3	.11
LTV	3	.05	4	.05	-2	.02	-8	.20
Pickup truck	16	.24	18	.12	none	.00	-24	.12
SUV	10	.31	18	.61	13	.39	15	.30
Van	-43	1.91	-37	1.44	-54	2.47	-50	2.08
CRASH TYPE								
Single-vehicle	11	.84	8	.34	10	.53	10	.42
Multivehicle	2	.04	-1	.00	-2	.04	-5	.19

CHAPTER 5

EFFECT OF REDESIGNED AIR BAGS ON CHILD PASSENGERS' OVERALL FATALITY RISK IN FRONTAL CRASHES

5.0   Summary

The fatality risk in frontal crashes of 0-12 year-old child passengers in the front seat decreased by a statistically significant 45 percent with sled-certified air bags (90 percent confidence bounds: 30 to 56 percent). Sled-certified air bags undid the harm wrought by the first generation of air bags for child passengers up to age 10, while preserving the life-saving benefits of those air bags for pre-teens age 11-12 years. Analyses are based on 1986-2004 Fatality Analysis Reporting System (FARS) data. For children up to age 10 years, approximately half the fatalities in frontal crashes with original air bags were "SCI fatalities": in low-to-moderate speed crashes and caused by the air bags. Thus, the results in this chapter are consistent with the 83 percent reduction of SCI fatalities of child passengers with sled-certified air bags, found in Chapter 2.

Even though sled-certified air bags mitigated much of the harm with original air bags, child passengers up to age 12 years should continue to travel in the back seat. When children travel in pickup trucks without a back seat, on-off switches for passenger air bags should be turned off. NHTSA will issue a research note comparing the fatality risk of child passengers in the back and front seats of vehicles with sled-certified air bags.

5.1   Background: effect of original air bags in frontal crashes

NHTSA's 1996 statistical analyses of the Fatality Analysis Reporting System (FARS) showed that early air bags increased the overall fatality risk of child passengers up to approximately age 10.[90] The cars and LTVs with early air bags have traveled many miles since 1996. By the end of 2004, they had accumulated a large portion of their lifetime vehicle exposure years. Let us update the analyses with the latest fatality counts to delineate the safety problem for child passengers.

Double-pair comparison is a good method for analysis of child passengers.[91] A large number of make-models were initially equipped with only a driver air bag for several years before 1994. At that time the passengers' seats changed from no air bags to air bag-equipped. The driver's seat stayed the same (air bag equipped). That allows double-pair comparison, with the driver as the control group. Because the number of child passengers in fatal crashes is small, at least relative to adults and teenagers, we need to use every possible FARS case. Double-pair comparison allows inclusion of the widest range of data: the ratio of front-seat passenger to driver fatalities is rather invariant over time, with vehicle age, or with the introduction of other safety devices that have similar effects for drivers and passengers.

---

[90] Kahane, C.J., *Fatality Reduction by Air Bags*, NHTSA Technical Report No. DOT HS 808 470, Washington, 1996, pp. 44-49.

[91] Evans, L., "Double Pair Comparison - A New Method to Determine How Occupant Characteristics Affect Fatality Risk in Traffic Crashes," *Accident Analysis and Prevention*, Vol. 18, June 1986, pp. 217-227.

The effect of air bags is analyzed in frontal impacts: in FARS, vehicles with the initial <u>or</u> principal impact point between 11:00 and 1:00, excluding first-event rollovers and non-collisions.[92] The analysis is based on CY 1986-2004 FARS data and includes all vehicles:

- Equipped with:
    - An air bag for the driver,
    - An air bag that is not sled-certified, or no air bag at all for the passenger, and
    - 3-point belts (not 2-point automatic belts) for the driver and right-front (RF) passenger.
- Occupied by at least one child passenger age 0-12 years in the front seat, and by a driver.[93]
- The child passenger, the driver, or possibly both died.

The fatalities tabulate as follows:

RF Passenger Seat	Driver's Seat	Child Passenger Fatalities	Driver Fatalities	Child/Driver Risk Ratio
No air bag	Air-bag equipped	270	297	.909
Original air bag	Air-bag equipped	348	261	1.333

Child passengers of cars equipped with air bags experienced a 47 percent increase of frontal fatalities:

$$1 - [(348/261) / (270/297)] = -.4667$$

The fatality increase is statistically significant, as evidenced by a Chi-square (χ^2) of 10.68 for the 2x2 table. χ^2 must exceed 3.84 for statistical significance at the two-sided .05 level. The 95

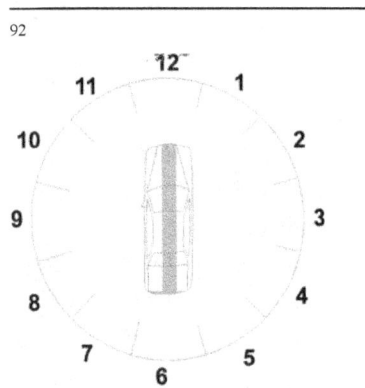

---

[92]

[93] The child passenger may be in the right-front, center-front or other/unknown front seat. In the few vehicles with two or more child passengers in the front seat, one double-pair case is created for each child, using the same driver record.

percent confidence bounds for the increase range from 17 to 85 percent.[94] However, Table 5-1 shows that the effect of air bags varies considerably with the age of the child.

TABLE 5-1

EFFECT OF ORIGINAL AIR BAGS ON FRONTAL FATALITIES
BY AGE OF CHILD PASSENGER
(1986-2004 FARS, as evidenced by ratio of child to driver fatalities – double pair comparison)

Passenger Seat	Child Fatalities	Driver Fatalities	Child/Driver Risk Ratio	Reduction (%)	χ^2
INFANTS AGE < 1 YEAR					
No air bag	30	27	1.482		
Original air bag	41	9	4.556	– 208	6.68
TODDLERS AGE 1 – 5 YEARS					
No air bag	93	95	.979		
Original air bag	164	73	2.247	– 129	17.07
CHILDREN AGE 6 – 10 YEARS					
No air bag	95	130	.731		
Original air bag	113	113	1.000	– 37	2.74
PRE-TEENS AGE 11 – 12 YEARS					
No air bag	42	45	.933		
Original air bag	30	66	.455	51	5.54

First-generation air bags increased fatality risk in frontal crashes by more than triple for infants, more than double for toddlers age 1-5 years and by 37 percent for children age 6-10 years. But the air bags were quite effective for pre-teens age 11-12 years, reducing fatality risk by 51 percent. The χ^2 statistics for the four 2x2 tables, shown in the last column, are all significant at the two-sided .05 level except for the analysis of children age 6-10 years. The findings are not surprising, for Table 1-2 showed a parallel variation in the proportion of frontal fatalities that took place in low-speed crashes and were caused by contact with air bags ("SCI fatalities"): 64 percent for infants, 51 percent for toddlers, 32 percent at age 6-10 years, but only 6 percent at age 11-12 years. Most of these children would have survived the crash, if not for the air bag. At

---

[94] Computed by the RELRISK option of the FREQ procedure in SAS® software.

age 11-12 years, original air bags only added a small proportion of SCI cases while preventing a larger number of fatalities that would have occurred without the bags.

Table 5-2 additionally compares unrestrained and restrained children in the two intermediate age groups. The effect for unrestrained toddlers is almost as drastic as for infants, a statistically significant, near threefold increase in fatality risk. Results for restrained toddlers and unrestrained children age 6-10 years are similar: fatality risk almost double, a statistically significant increase. But for belted children age 6-10 years, air bags had a negligible, non-significant effect.

TABLE 5-2

EFFECT OF ORIGINAL AIR BAGS ON FRONTAL FATALITIES
BY AGE AND RESTRAINT USE OF CHILD PASSENGER
(1986-2004 FARS, as evidenced by ratio of child to driver fatalities – double pair comparison)

Passenger Seat	Child Fatalities	Driver Fatalities	Child/Driver Risk Ratio	Reduction (%)	χ^2
UNRESTRAINED TODDLERS AGE 1 – 5 YEARS					
No air bag	43	35	1.229		
Original air bag	109	32	3.406	– 177	11.63
TODDLERS AGE 1 – 5 YEARS IN CHILD SAFETY SEAT OR SAFETY BELTS					
No air bag	42	55	.764		
Original air bag	47	32	1.469	– 92	4.57
UNRESTRAINED CHILDREN AGE 6 – 10 YEARS					
No air bag	43	48	.896		
Original air bag	54	31	1.742	– 94	4.71
BELTED CHILDREN AGE 6 – 10 YEARS					
No air bag	48	73	.658		
Original air bag	54	79	.684	– 4	.02

In other words, the objective of redesigned air bags for child passengers up to age 10 years is to undo, as much as possible, the harm wrought by the first generation of air bags. But for pre-teens age 11-12 years, the goal is to preserve the impressive fatality reductions of early air bags, as for drivers and adult/teenage passengers.

5.2   Double-pair comparison analyses of sled-certified air bags

The same analysis method can estimate the effect of sled-certified air bags, relative to original air bags for child passengers in frontal crashes. In fact, all we need to do is add a line for sled-certified air bags to each segment of the preceding tables.

Earlier, Sections 3.1 and 4.1 explained that double-pair comparison could not be used for analyses of drivers or adult passengers because driver and passenger air bags were sled-certified at the same time or at most a year apart, violating the requirement that one seat position stay the same while the other is modified. But circumstances of the analysis have changed. With drivers and adult passengers, the effect of redesigning air bags, if there was any at all, was anticipated to be small. Anything that could bias the results even by a few percent was to be strenuously avoided. Plenty of data were available.

With child passengers, we can expect a substantial benefit for redesigned air bags based on the results of the SCI analyses in Section 2.1, among other things. Chapter 3 showed that the effect of redesigning driver air bags was zero (point estimate) or, at most, negligible (interval estimate) relative to the likely effect for child passengers. Although driver air bags were sled-certified at the same time as passenger air bags, they may be regarded as "unchanged" and a valid comparison group for the analysis of child passengers. Because the N of child passenger fatalities is low, we need the widest possible range of model years in the data. Double-pair comparison allows a wide range, because the ratio of child to driver fatalities is rather invariant over time or as a vehicle ages.

Unlike the two preceding chapters, there will be no parallel analyses limited to make-models where it is known that air bags were depowered, because it would be futile to narrow even further the already limited available data.

FARS data preparation: The analysis is based on FARS data from calendar years 1986-2004 and model years 1985-2002. All make-models are included except a very few known to have up-powered when they sled-certified on the passenger side. NHTSA has fairly complete tables of when each make-model's air bags were initially sled-certified.[95] For make-models not in those tables, it is certain they were not sled-certified through model year 1997; we will assume they are sled-certified from 1999 onwards and exclude model year 1998.[96] As in Section 5.1, every vehicle in the analysis has to be equipped with driver air bags as well as three-point safety belts at the driver's and right-front seat (no 2-point automatic belts). Pickup trucks with on-off switches for the passenger air bag are not included for now, but will be analyzed separately in Table 5-6. None of the previous chapters' additional limitations are placed on the range of

---

[95] *National Automotive Sampling System – Crashworthiness Data System – 2000 Coding and Editing Manual*, NHTSA, Washington, 2000, pp. 721-733, access from http://www-nrd.nhtsa.dot.gov/departments/nrd-30/ncsa/AvailInf.html .

[96] Similarly, if a model sled-certified in the middle of a model year, we will identify the sled-certified vehicles based on the VIN if possible and exclude that model year otherwise.

73

model years.[97] Child passengers in the right-front, center-front and other/unknown front seat are included.

Basic FARS analysis of sled-certified air bags: Table 5-3 presents the entire dataset for child passengers. With original air bags, 348 children and 261 accompanying drivers were killed in frontal crashes, a risk ratio of 1.333. With sled-certified air bags, there were 132 child and 179 driver fatalities, a risk ratio of 0.737. That is a 45 percent reduction of children's fatality risk with sled-certified air bags as compared to original air bags, and it is statistically significant, as evidenced by a χ^2 of 17.83. χ^2 must exceed 3.84 for statistical significance at the two-sided .05 level, but just 2.71 for significance at the one-sided .05 level. The 95 percent confidence bounds for the reduction range from 27 to 58 percent; 90 percent confidence bounds range from 30 to 56 percent.[98]

NHTSA evaluations of safety standards customarily employ the more lenient one-sided test and 90 percent confidence bounds, because there is a clear expectation that the effect of the standard will either be in the "right" direction (saving lives) or at worst zero. It is unlikely to be negative. That strategy also makes sense for the remainder of this chapter, because we clearly expect redesigned air bags to do less harm to children than original air bags, certainly not to do more. (But Chapters 3 and 4 used the more stringent two-sided test and 95 percent confidence bounds because, for drivers and adult passengers, redesigned air bags might conceivably increase or reduce fatality risk.)

TABLE 5-3

EFFECT OF SLED-CERTIFIED AIR BAGS ON FRONTAL FATALITIES
OF FRONT-SEAT CHILD PASSENGERS AGE 0-12 YEARS
(1986-2004 FARS, as evidenced by ratio of child to driver fatalities – double pair comparison)

Passenger Seat	Child Fatalities	Driver Fatalities	Child/Driver Risk Ratio	Reduction from Original (%)	χ^2
No air bag	*270*	*297*	*.909*		
Original air bag	348	261	1.333		
Sled-certified air bag	132	179	.737	45	17.83

In fact, the fatality risk ratio with sled-certified air bags (.737) is actually lower than with no passenger air bag (.909). However, the analysis in Table 5-3 may not be the most accurate way

---

[97] Namely, limiting the data to ± 3 model years; excluding dual-stage air bags or a change in pretensioners (second-order effects); excluding a change in side impact protection or side air bags (double-pair comparison analysis is limited to frontal impacts); excluding a change in ESC (equally beneficial for driver and passenger; thus, will not affect double-pair comparison); "balancing" the sample.

[98] 95% confidence bounds are computed by the RELRISK option of the FREQ procedure in SAS® software. For 90% confidence bounds take ±1.645 rather than ±1.96 standard deviations of the log-odds ratio.

to estimate the benefit of redesigned air bags. As more parents moved the youngest children to the back seat (see Section 1.5), the median age of children remaining in the front seat increased. As a child grows from age 0 to 12 years, probability of surviving a similar physical insult steadily increases[99]; the low fatality risk for the sled-certified air bags partly reflects a higher share of older children among the front-seat passengers. In addition, the effect of sled-certified air bags is likely not uniform at all ages. For these reasons, each age group should be analyzed separately, to the extent allowed by the available data.

TABLE 5-4

EFFECT OF SLED-CERTIFIED AIR BAGS ON FRONTAL FATALITIES
BY AGE OF CHILD PASSENGER
(1986-2004 FARS, as evidenced by ratio of child to driver fatalities – double pair comparison)

Passenger Seat	Child Fatalities	Driver Fatalities	Child/Driver Risk Ratio	Reduction from Original (%)	χ^2
		INFANTS AGE < 1 YEAR			
No air bag	*30*	*27*	*1.482*		
Original air bag	41	9	4.556		
Sled-certified air bag	12	1	12.000	$- 163^{100}$	
		TODDLERS AGE 1 – 5 YEARS			
No air bag	*93*	*95*	*.979*		
Original air bag	164	73	2.247		
Sled-certified air bag	47	50	.940	58	12.73
		CHILDREN AGE 6 – 10 YEARS			
No air bag	*95*	*130*	*.731*		
Original air bag	113	113	1.000		
Sled-certified air bag	50	73	.685	31	2.80
		PRE-TEENS AGE 11 – 12 YEARS			
No air bag	*42*	*45*	*.933*		
Original air bag	30	66	.455		
Sled-certified air bag	23	55	.418	8	.08

[99] Evans, L., *Traffic Safety and the Driver*, Van Nostrand Reinhold, New York, 1991, pp. 25-28.
[100] Obviously not a statistically meaningful result, since it is based on just 1 driver fatality. Also, χ^2 is not meaningful because of the small cell.

Effect by child's age group: Table 5-4 computes fatality risk ratios for four age groups of children. Crash data for infants less than a year old are insufficient for a statistically meaningful analysis. Because the vast majority of infants now ride in the back seat, there is only a single case of a driver fatality in a vehicle with an infant in the front seat. Section 5.3 will analyze the infant fatality rate per billion vehicle registration years, a method that does not use any control group such as driver fatalities.

What is notable about the 12 infant fatalities in sled-certified vehicles is that FARS did not report any of them in a child safety seat. Ten were unrestrained. Restraint use was unknown for the other two. An unrestrained infant typically rides on the lap or in the arms of another passenger. That is a terribly exposed position if an air bag deploys, likely as dangerous as riding in a rear-facing safety seat.

The second segment of Table 5-4 shows that sled-certified air bags greatly reduced the risk ratio for toddlers age 1-5 years, to 0.940 from 2.247 with original air bags. This 58 percent fatality reduction is statistically significant, as evidenced by a χ^2 of 12.73. In fact, the .940 risk ratio with sled-certified air bags is almost identical to the .979 without any passenger air bag. In other words, the redesign of air bags undid the harm of first-generation air bags and essentially restored the *status quo ante* without air bags.

The results for children are 6-10 years are directionally similar. The fatality reduction for sled-certified air bags relative to original air bags is 31 percent, statistically significant at the one-sided .05 level. The .685 risk ratio with sled-certified air bags is slightly, but not significantly lower than the .731 without an air bag ($\chi^2 = 0.08$). Here, too, sled-certified air bags apparently restored the situation before air bags.

For pre-teens age 11-12 years, the .418 fatality risk with sled-certified air bags is 8 percent lower than the .455 with original air bags. The reduction is not statistically significant. But it is significantly lower than the .933 with no air bags ($\chi^2 = 6.08$). In other words, redesigned air bags entirely preserved the great life-saving benefits of first-generation driver air bags for pre-teens.

The effectiveness of sled-certified air bags in Tables 5-3 and 5-4, except for infants, is consistent with, and may even exceed the benefits implied by the analysis of SCI fatalities (caused by air bags in low-to-moderate speed crashes) in Sections 1.3 and 2.1. Table 1-1 suggested that 42 percent of frontal fatalities to all children age 0-12 years with original air bags were SCI fatalities; Table 1-2 found 51 percent of toddlers' frontal fatalities, 32 percent of children age 6-10 years, but only 6 percent of pre-teens' frontal fatalities were SCI cases. Table 2-1b indicated an 83 percent reduction of SCI fatalities for sled-certified air bags. That effect alone would account for an 83 x .42 = 35 percent reduction of frontal fatalities for all children age 0-12 years, as well as an 83 x .51 = 42 percent reduction for toddlers, 83 x .32 = 27 percent for age 6-10 years and 83 x .06 = 5 percent reduction for pre-teens. The corresponding estimates in Tables 5-3 and 5-4 are each a bit higher: 45, 58, 31 and 8 percent, respectively. They suggest that redesigned air bags may have prevented not only SCI fatalities but also some fatal contacts with air bags in frontal crashes that were too severe to be SCI cases but still potentially survivable.

Effect by age group and restraint use: Table 5-5 compares unrestrained and restrained children in the two intermediate age groups. There are not enough data in the infant and pre-teen groups for separate analyses of unrestrained and restrained children.

TABLE 5-5

EFFECT OF SLED-CERTIFIED AIR BAGS ON FRONTAL FATALITIES
BY AGE AND RESTRAINT USE OF CHILD PASSENGER
(1986-2004 FARS, as evidenced by ratio of child to driver fatalities – double pair comparison)

Passenger Seat	Child Fatalities	Driver Fatalities	Child/Driver Risk Ratio	Reduction from Original (%)	χ^2
UNRESTRAINED TODDLERS AGE 1 – 5 YEARS					
No air bag	*43*	*35*	*1.229*		
Original air bag	109	32	3.406		
Sled-certified air bag	28	16	1.750	49	3.26
TODDLERS AGE 1 – 5 YEARS IN CHILD SAFETY SEAT OR SAFETY BELTS					
No air bag	*42*	*55*	*.764*		
Original air bag	47	32	1.469		
Sled-certified air bag	17	32	.531	64	7.44
UNRESTRAINED CHILDREN AGE 6 – 10 YEARS					
No air bag	*43*	*48*	*.896*		
Original air bag	54	31	1.742		
Sled-certified air bag	17	18	.944	46	2.30
BELTED CHILDREN AGE 6 – 10 YEARS					
No air bag	*48*	*73*	*.658*		
Original air bag	54	79	.684		
Sled-certified air bag	31	51	.608	11	.17

Fatality risk in frontal crashes was lower with sled-certified air bags than with original air bags in all four cohorts: unrestrained as well as restrained children. The reductions were significant at the one-sided .05 level for unrestrained and restrained toddlers. Toddlers in child safety seats or belts experienced the largest reduction, 64 percent. Restrained children in both age groups had lower fatality risk with sled-certified air bags than with no air bags at all, but unrestrained children had at least somewhat higher fatality risk, even with sled-certified air bags, than with no air bags.

Effect by vehicle type: Table 5-6 computes the effect of sled-certified air bags for all children age 0-12 years, separately for passenger cars, pickup trucks, SUVs and vans.

TABLE 5-6

EFFECT OF SLED-CERTIFIED AIR BAGS ON FRONTAL FATALITIES
OF FRONT-SEAT CHILD PASSENGERS AGE 0-12 YEARS, BY VEHICLE TYPE
(1986-2004 FARS, as evidenced by ratio of child to driver fatalities – double pair comparison)

Passenger Seat	Child Fatalities	Driver Fatalities	Child/Driver Risk Ratio	Reduction from Original (%)	χ^2
PASSENGER CARS					
No air bag	*170*	*186*	*.914*		
Original air bag	282	227	1.242		
Sled-certified air bag	95	109	.872	30	4.56
PICKUP TRUCKS					
No air bag	*43*	*44*	*.977*		
Original + on-off switch	22	10	2.200		
Sled-cert. + on-off switch	46	46	1.000	55	3.37
SUVs					
No air bag	*18*	*20*	*.900*		
Original air bag	19	16	1.188		
Sled-certified air bag	23	37	.622	48	2.28
VANS					
No air bag	*39*	*47*	*.830*		
Original air bag	47	18	2.611		
Sled-certified air bag	8	27	.296	89	22.48

The cases for cars, SUVs and vans are all part of the dataset used in Table 5-3. Data for pickup trucks with passenger air bags, on the other hand, are limited to trucks with factory-equipped on-off switches, and were not part of any previous table. Table 5-6 excludes pickup trucks without on-off switches, such as full crew cabs. We do not want to compare air bags that are "on" all the time to air bags that can be turned off.

Fatality reductions for sled-certified relative to original air bags are substantial in all vehicle types, ranging from 30 to 89 percent. Three of the four reductions are significant at the one-sided .05 level. Three of the four risk ratios with sled-certified air bags are lower than the corresponding ratio without passenger air bags, and the fourth is about the same.

The 30 percent fatality reduction in passenger cars, although statistically significant is the lowest among the four vehicle types. With lower mass and smaller size than LTVs, cars presumably experience a substantial proportion of high-severity impacts with other, heavier vehicles. Those are not the type of crashes where original air bags caused extra harm and redesigned air bags prevented it.

The 55 percent fatality reduction with pickup trucks is noteworthy, considering all the trucks were equipped with on-off switches. If the switches had all been correctly used – i.e., turned off for child passengers – the fatality ratios for original and redesigned air bags should both have been the same as with no air bags. The high ratio of child to driver fatalities with original air bags hints that switches must have been left on for many of the children in these crashes. The substantial reduction with sled-certified air bags may indicate a combination of less aggressive deployments and better use of the switches in the latest vehicles.

The substantial effect in SUVs, no different from the other vehicle types, is reassuring given the previous findings by the Children's Hospital of Philadelphia that redesigned air bags were not reducing injury risk in SUVs (see Section 1.6).

The 89 percent fatality reduction in vans is remarkable. The fatality rate with original air bags (2.611) was the highest of the four vehicle types, but after sled certification, vans have the lowest rate (0.296). Here is more evidence that manufacturers may have made an extra effort to reduce the aggressiveness of passenger air bags in vans, greatly reducing their risk for child passengers (see also Sections 2.1 and 4.4).

Effect by crash type: Table 5-7 compares the effect of sled-certified air bags for all children age 0-12 years in single-vehicle and multivehicle crashes. Fatality reduction, relative to original air bags is a non-significant 21 percent in single-vehicle crashes and a statistically significant 51 percent in multivehicle crashes. Moreover, the fatality risk with original air bags was about the same as with no air bags at all in single-vehicle crashes, but much higher in multivehicle crashes.

The explanation for the difference is simple: 83 percent of the SCI fatalities of child passengers were in multivehicle crashes. An especially well-known SCI case involved running over a curb in a parking lot, but actually the vast majority of child SCI cases were collisions of minor to moderate severity with other vehicles.

TABLE 5-7

EFFECT OF SLED-CERTIFIED AIR BAGS ON FRONTAL FATALITIES
OF FRONT-SEAT CHILD PASSENGERS AGE 0-12 YEARS, BY CRASH TYPE
(1986-2004 FARS, as evidenced by ratio of child to driver fatalities – double pair comparison)

Passenger Seat	Child Fatalities	Driver Fatalities	Child/Driver Risk Ratio	Reduction from Original (%)	χ^2
SINGLE-VEHICLE CRASHES					
No air bag	*77*	*90*	*.856*		
Original air bag	76	95	.800		
Sled-certified air bag	41	65	.631	21	.89
MULTIVEHICLE CRASHES					
No air bag	*193*	*207*	*.932*		
Original air bag	272	166	1.639		
Sled-certified air bag	91	114	.798	51	17.82

5.3   Frontal fatalities per billion vehicle registration years

A second analysis method is to compare front-seat child passengers' frontal fatality rates per billion vehicle registration years before vs. after sled certification. The analysis is based on FARS data and R.L. Polk's National Vehicle Population Profile (NVPP). NVPP specifies the number of vehicles of any make-model and model year that were registered as of July 1 of any calendar year: a surrogate for the number of registration years (exposure) they accumulated in that calendar year.

Many definitions are the same as in Section 5.2, namely:

- Includes "sled-certified" and "original" passenger air bags (same definitions).

- Includes all make-models except the few known to have up-powered.

- Includes all model years through 2002.

- Excludes pickup trucks with on-off switches.

- Excludes vehicles with motorized automatic belts for the passenger seat.

- Includes child passengers in the right-front, center-front and other/unknown front seat.

- "Frontal" crashes have initial <u>or</u> principal impact between 11:00 and 1:00 (excluding first-event rollovers and non-collisions).

As in Chapter 2, the analysis excludes vehicles of the "next" model year with MY = CY + 1.[101] Also, if a make-model was sled-certified in mid-model year, or if it had a mix of vehicles with and without passenger air bags in the same year, that model year needs to be excluded from the analysis because NVPP does not distinguish between vehicles without passenger air bags, with original air bags, or with sled-certified air bags within the same model year.

Finally, one additional cohort of data must be excluded. Fatalities of front-seat child passengers declined over time because more and more parents placed children in the back seat in response to public information about risk with air bags (see Section 1.5). Double-pair comparison analyses automatically adjusted for that (because fatalities to drivers accompanying the child passengers would decline by the same proportion). Fatalities per billion vehicle years, on the other hand, would decrease if occupancy decreases.[102] Children's occupancy of the front seat primarily changed over time – i.e., by calendar year, as the public became more aware of risks with air bags. Therefore, it is necessary to compare fatality rates with original and redesigned air bags in the same calendar years: specifically, calendar years 1998 and later, the only calendar years when there will be vehicles with redesigned as well as original air bags (after excluding the few MY 1998 vehicles already on the road in CY 1997). Calendar years 1997 and earlier need to be excluded because the sample consists exclusively of vehicles with original air bags.

Table 5-8 shows the rates of fatalities in frontal crashes, per billion registration years, for child passengers age 0-12 years riding in the front seat.

TABLE 5-8

FRONT-SEAT CHILD PASSENGERS AGE 0-12 YEARS, FRONTAL FATALITIES
PER BILLION VEHICLE YEARS, BEFORE AND AFTER SLED CERTIFICATION
(1998-2004 FARS and NVPP)

Passenger Seat	Frontal (11-1:00) Fatalities	Vehicle Years	Frontal Fatalities per Billion Years	Reduction (%)
Original air bag	236	237,914,490	991.95	
Sled-certified air bag	120	268,904,699	446.26	55

The fatality rate with sled-certified air bags (446) is 55 percent lower than with original air bags (992). That reduction is fairly close to the 45 percent fatality reduction found in the double-pair comparison analysis (Table 5-3); in fact, it is within the 90 percent confidence bounds of that

---

[101] NVPP says how many vehicles of model year X were registered on July 1 of calendar year Y. Early vehicles of the "next" model year (X = Y + 1) are entirely omitted in some years of NVPP, and even when included the count of vehicles registered on July 1 is a poor estimate of the registration years they accumulated. Strictly speaking, even for vehicles of the "current" model year (X = Y), the registration count for July 1 is not that accurate an estimate of the registration years accumulated by that MY in that CY, but we will accept it here because deleting all SCI cases with MY = CY would be too great a loss of data.

[102] That was not a problem in the FARS-NVPP analyses of Chapters 3 and 4 because the driver's seat is always occupied and, as far as we know, adult passengers did not move to the back seat in substantial numbers.

earlier estimate (30 to 56 percent). The most likely explanation for the somewhat higher reduction with this method is that limiting the data to calendar years 1998-2004 does not fully correct for differences in vehicle occupancy; perhaps, in any given calendar year, owners of newer vehicles are better informed and more diligent about moving children to the back seat.[103] In other words, we continue to believe that Table 5-3 provides the best estimate of overall fatality reduction with sled-certified air bags, 45 percent, whereas Table 5-8 produces a fairly comparable but somewhat higher estimate, 55 percent.

However, as discussed in Section 5.2, it is important to compute effectiveness separately for various age groups of children, because it can vary substantially. Double-pair comparison did not provide a satisfactory analysis for infants less than a year old because there were so few fatalities to drivers accompanying the infants (see Table 5-4). The present method works better, at least for that group, because it does not require information on driver fatalities. Table 5-9 computes fatality rates for four age groups of children.[104]

Infants' fatality rate is 109 with original air bags and 45 with sled-certified air bags, a 59 percent reduction. That is slightly higher than the 55 percent overall reduction for children age 0-12 years in Table 5-8. Even taking into account that this method produces somewhat high estimates, it still suggests that redesigned air bags are approximately as beneficial for infants as for toddlers and 6-10 year-old children.

The second segment of Table 5-9 shows a 67 percent fatality reduction with sled-certified air bags for toddlers age 1-5 years. It is not too different, although somewhat higher than the 58 percent fatality reduction, based on double-pair comparison, in Table 5-4. Similarly, the 48 and 16 percent reductions in the fatality rates for children age 6-10 and 11-12 years, respectively, moderately exceed the corresponding estimates with double-pair comparison: 31 and 8 percent.

---

[103] Even when the data in Table 5-8 are further limited to CY 1999-2004, the fatality reduction is nearly the same, 54 percent, and likewise if they are limited to CY 2000-2004.
[104] Fatality rates should not be compared directly across age groups. The four age groups include ranges of children's age of 1, 5, 5 and 2 years, respectively. The fatality rate increases more or less proportionately to the range of children's ages included.

TABLE 5-9

FRONT-SEAT CHILD PASSENGERS AGE 0-12 YEARS, FRONTAL FATALITIES PER BILLION VEHICLE YEARS, BEFORE AND AFTER SLED CERTIFICATION BY AGE OF THE CHILD PASSENGER
(1998-2004 FARS and NVPP)

Passenger Seat	Frontal (11-1:00) Fatalities	Vehicle Years	Frontal Fatalities per Billion Years	Reduction (%)
INFANTS AGE < 1 YEAR				
Original air bag	26	237,914,490	109.28	
Sled-certified air bag	12	268,904,699	44.63	59
TODDLERS AGE 1 – 5 YEARS				
Original air bag	109	237,914,490	458.15	
Sled-certified air bag	41	268,904,699	152.47	67
CHILDREN AGE 6 – 10 YEARS				
Original air bag	79	237,914,490	332.05	
Sled-certified air bag	46	268,904,699	171.06	48
PRE-TEENS AGE 11 – 12 YEARS				
Original air bag	22	237,914,490	92.47	
Sled-certified air bag	21	268,904,699	78.10	16

REFERENCES

Air Bag & Seat Belt Safety Campaign, Air Bag & Seat Belt Safety Tips. National Safety Council, Chicago, November 17-30, 2003. (http://www.nsc.org/partners/safetips.htm).

Air Bag Test Data – NHTSA Research & Development. (http://www-nrd.nhtsa.dot.gov/departments/nrd-11/airbags/abgdb/).

Air Bags & On-Off Switches. NHTSA Publication No. DOT HS 808 629, Washington, 1997.

Aldman B, Anderson A, and Saxmark O. "Possible Effects of Air Bag Inflation on a Standing Child," *Proceedings of the 18th Annual Conference of the American Association for Automotive Medicine*, Morton Grove, IL, 1974.

Arbogast, K.B., Durbin, D.R., Kallan, M.J., and Winston, F.K. "Effect of Vehicle Type on the Performance of Second Generation Air Bags for Child Occupants," *47th Annual Proceeding – Association for the Advancement of Automotive Medicine*, Barrington, IL, 2003, pp. 85-99.

Automobile Occupant Crash Protection Progress Report No. 3. NHTSA Report No. DOT HS 805 474, Washington, 1980.

Braver, E.R., Kyrychenko, S.Y., and Ferguson, S.A. "Driver Mortality in Frontal Crashes: Comparison of Newer and Older Air Bag Designs," *Traffic Injury Prevention*, Vol. 6, March 2005, pp. 24-30.

Buckle-Up America Child Passenger Safety Week, February 10-16, 2002, Talking Points, NHTSA, (http://www.nhtsa.dot.gov/people/injury/airbags/buckleplan/ CPS%20Week%20Planner_files/talking1.html).

Code of Federal Regulations, Title 49. Government Printing Office, Washington, 2005.

Evaluation of the Effectiveness of Occupant Protection, Interim Report. NHTSA Report No. DOT HS 807 843, 1992.

Evans, L. "Double Pair Comparison - A New Method to Determine How Occupant Characteristics Affect Fatality Risk in Traffic Crashes," *Accident Analysis and Prevention*, Vol. 18, June 1986, pp. 217-227.

_____. *Traffic Safety and the Driver.* Van Nostrand Reinhold, New York, 1991.

Federal Register Notices:

> 58 (September 2, 1993): 46551, Final Rule amending FMVSS 208 to require warning labels that rear-facing safety seats should never be placed in the front seat.

> 60 (May 23, 1995): 27233, Final Rule amending FMVSS 208 to allow on-off switches for passenger air bags in pickup trucks and some other vehicles.

60 (November 9, 1995): 56554, requests comments and announces public meeting to discuss technological changes to reduce adverse effects of air bags.

61 (August 6, 1996): 40784, NPRM to reduce adverse effects of air bags.

61 (November 27, 1996): 60206, Final Rule amending FMVSS 208 to require warning labels on air bags' risk to children up to age 12 and to specify that the back seat is the safest place for children.

62 (January 6, 1997): 807, NPRM to modify the FMVSS 208 test to permit air bags that deploy less forcefully..

62 (March 19, 1997): 12960, Final Rule modifying the FMVSS 208 test to permit air bags that deploy less forcefully.

62 (November 21, 1997): 62406, sets up procedures enabling the public to obtain aftermarket on-off switches for air bags.

65 (May 12, 2000): 30679, Final Rule amending FMVSS 208 to require advanced air bags.

68 (August 6, 2003): 46539, NPRM to require a 35 mph test with 5th percentile female dummies for advanced air bags.

Ferguson, S.A., Schneider, L., Segui-Gomez, M., Arbogast, K., Augenstein, J., and Digges, K.H. "The Blue Ribbon Panel on Depowered and Advanced Airbags – Status Report on Airbag Performance," *47th Annual Proceeding – Association for the Advancement of Automotive Medicine*, Barrington, IL, 2003, pp. 79-102.

Fifth/Sixth Report to Congress – Effectiveness of Occupant Protection Systems and Their Use. NHTSA Report No. DOT HS 809 442, Washington, 2001.

Final Economic Assessment, FMVSS 208, Advanced Air Bags. NHTSA Docket No. NHTSA-2000-7013-2, 2000.

Final Regulatory Evaluation, Depowering. NHTSA Docket No. NHTSA-1997-2817-002, 1997.

Fourth Report to Congress – Effectiveness of Occupant Protection Systems and Their Use. NHTSA Report No. DOT HS 808 919, Washington, 1999.

Healy, J.R., and O'Donnell, J. "Deadly Air Bags," *USA Today*, July 8, 1996.

Hinch, J., Hollowell, W.T., Kanianthra, J., Evans, W.D., Klein, T., Longthorne, A., Ratchford, S., Morris, J., and Subramanian, R. *Air Bag Technology in Light Passenger Vehicles.* National Highway Traffic Safety Administration, Washington, 2001. (http://www-nrd.nhtsa.dot.gov/departments/nrd-11/NRDpubs.html).

Kahane, C.J. *Fatality Reduction by Air Bags: Analyses of Accident Data through Early 1996.* NHTSA Technical Report No. DOT HS 808 470, Washington, 1996.

_____. *Lives Saved by the Federal Motor Vehicle Safety Standards and Other Vehicle Safety Technologies, 1960-2002.* NHTSA Technical Report No. DOT HS 809 833, Washington, 2004.

Kindelberger, J. and Starnes, M. *Moving Children from the Front Seat to the Back Seat: The Influence of Child Safety Campaigns.* NHTSA Research Note No. DOT HS 809 698, Washington, 2003.

Kindelberger, J.C., Chidester, A.B., and Ferguson, E. "Air Bag Crash Investigations," *Proceedings 18th International Technical Conference on the Enhanced Safety of Vehicles.* NHTSA Report No. DOT HS 809 543, Washington, 2003, Paper No. 299.

Montalvo, F., Bryant, R., and Mertz, H. *Possible Positions and Postures of Unrestrained Front-Seat Children at the Instant of Collision.* Paper No. 826045, Society of Automotive Engineers, Warrendale, PA, 1982.

Morgan, C. *Results of the Survey on the Use of Passenger Air Bag On-Off Switches.* NHTSA Technical Report No. DOT HS 809 689, Washington, 2003.

National Automotive Sampling System – Crashworthiness Data System – 2000 Coding and Editing Manual, NHTSA, Washington, 2000. (http://www-nrd.nhtsa.dot.gov/departments/nrd-30/ncsa/AvailInf.html).

NHTSA Permits Air Bag Switch to Prevent Injury to Infants in Rear-Facing Safety Seats. Press Release No. NHTSA 30-95, U. S. Department of Transportation, Office of the Assistant Secretary for Public Affairs, Washington, 1995.

NHTSA Warns Parents about Child Safety Seat Use in Cars with Air Bags. Press Release No. NHTSA 60-91, U. S. Department of Transportation, Office of the Assistant Secretary for Public Affairs, Washington, 1991.

Observed Safety Belt Use in 1996. NHTSA (NCSA) Research Note, Washington, 1997.

Patrick, L.M. and Nyquist, G.W. *Airbag Effects on the Out-of-Position Child.* Paper No. 720442, Society of Automotive Engineers, Warrendale, PA, 1972.

Safety Agency Issues Warning on Air Bag Danger to Children. Press Release No. NHTSA 72-95, U. S. Department of Transportation, Office of the Assistant Secretary for Public Affairs, Washington, 1995.

SAS/STAT® User's Guide, Vol. 1, Version 6, 4th Ed. SAS Institute, Cary, NC, 1990.

Secretary Peña Announces Government/Industry Coalition for Air Bag Safety. Press Release No. NHTSA 24-96, U. S. Department of Transportation, Office of the Assistant Secretary for Public Affairs, Washington, 1996.

Special Crash Investigations – NHTSA National Center for Statistics and Analysis. (http://www-nrd.nhtsa.dot.gov/departments/nrd-30/ncsa/sci.html).

Status Report: On the Issue of Testing Air Bag-Equipped Vehicles with and without Belt Restraints at Different Speeds. NHTSA Docket No. NHTSA-1996-1772-2, 1995.

Third Report to Congress – Effectiveness of Occupant Protection Systems and Their Use. NHTSA Report No. DOT HS 808 537, Washington, 1996.

Traffic Safety Facts – 2004 Data – Occupant Protection. NHTSA Report No. DOT HS 809 909, Washington, 2005.

Walz, M.C. *NCAP Test Improvements with Pretensioners and Load Limiters.* NHTSA Technical Report No. DOT HS 809 562, Washington, 2003.